Second Edition

CHEMISTRY

in the NATIONAL SCIENCE EDUCATION STANDARDS

Models for Meaningful Learning in the High School Chemistry Classroom

Edited by Stacey Lowery Bretz

ACKNOWLEDGMENTS

A special thanks to the authors of the first edition of the *Chemistry in the National Science Education Standards*: Stanley Pine, Henry Heikkinen, Sylvia Ware, Michael Tinnesand, Diane Bunce, Jerry Bell, Patricia Smith, Bonnie Brunkhorst, Ann Benbow, Conrad Stanitski, Mary Virginia Orna, Kathryn Scantlebury, Dwaine Eubanks, Lucy Eubanks, and Ronald Archer.

The American Chemical Society would also like to thank the following:
The National Academies Press for permission to reprint material from the *National Science Education Standards* (© 1996, National Academy Press).

The *Journal of Chemical Education* and the Division of Chemical Education, Inc., for permission to reprint material from the *Journal of Chemical Education* (© 1993, Division of Chemical Education, Inc.).

John Wiley & Sons, Inc., for permission to reprint material from *Chemistry: A Guided Inquiry*, 3rd Edition (John Wiley & Sons, Inc., © 2006).

Wiley-Liss, Inc. a subsidiary of John Wiley & Sons for permission to reprint material from the *Journal of Research in Science Teaching* (© 1994 and © 1999 National Association for Research in Science Teaching).

A special thanks to Tanya Sharpe, John Eggebrecht, and Marcia Wilbur of the Advanced Placement Program at the College Board.

The production of this report has been in the following very capable hands:
Cornithia Harris, *Art Director*
Leona Kanaskie, *Copy Editor*
Pete Isikoff, *Program Assistant*
Terri Taylor, *Assistant Director, K–12 Science*
Michael Tinnesand, *Associate Director, Academic Programs*

ISBN No. 978-0-8412-6991-0

CONTENTS

Teaching and Learning High School Chemistry

Stacey Lowery Bretz is *Professor of Chemistry at Miami University in Oxford, OH. Her research expertise includes assessment of student learning, applications of cognitive science and learning theory to chemistry education, and how children learn chemistry. Previous initiatives have included curriculum reform in general chemistry and the professional development of high school chemistry teachers. Dr. Bretz chaired the Gordon Research Conference on CER in 2005 and serves on the Board of Trustees for the American Chemical Society Examinations Institute. Contact information: bretzsl@muohio.edu*

by Stacey Lowery Bretz

Editor, Chemistry in the National Science Education Standards, 2nd ed.

In 1996 the National Research Council published the *National Science Education Standards*. This document was the result of a long, collaborative process intended to specify what **students** should *know* and what students should *be able to do* after graduating from high school, what **teachers** should *know* about science and *be able to do* to teach science, and what **education systems** should *know* about educating future science teachers and *be able to do* to ensure students would learn science from these teachers. In conjunction with the American Association for the Advancement of Science's *Benchmarks for Science Literacy*, the NSES catalyzed conversations in each of the 50 states about what science was being taught, what science should be taught, and how teachers would assess the quality and quantity of student learning in science.

The Standards did not delineate content for chemistry. They did not specify content for physics, geology, botany, zoology, or microbiology, either. Rather the Standards were organized across physical science, earth science, life science, inquiry, and social and personal perspectives — prompting some chemists and chemistry teachers to ask, "where's the chemistry?"

However, the *NSES* certainly included the concepts important to chemistry because the American Chemical Society's Committee on Education (SOCED) actively participated in constructing the Standards. As outreach to high school chemistry teachers, the American Chemical Society (ACS) Education Division commissioned the first edition of this book, *Chemistry in the National Science Education Standards*. Since that first edition in 1996, much has changed — many state standards have been updated to reflect the NSES and *Benchmarks,* the Internet is ubiquitous now, technology has transformed how chemists gather and interpret data, and *No Child Left Behind* (*NCLB*) is the law of the land.

This 2nd edition of *Chemistry in the National Science Education Standards* responds to the changing landscape of teaching high school chemistry by providing updated models for meaningful learning. All chapters have been thoroughly updated and several new ones on technology, English language learners, student misconceptions, and learning research have been added. Each chapter contains recommended Web sites and additional readings, as well as contact information for the authors so you can get additional information on topics of interest. A brief description of each chapter is offered below to whet your appetite.

Who should use this book? Certainly high school chemistry teachers and administrators should find this book contains valuable and thought-provoking information. ***Each chapter has been coauthored by a high school chemistry teacher,*** *w*hich means that the concrete, practical examples contained in each chapter have already been classroom tested. University chemistry and science education faculty members, both of whom teach preservice teachers, will find this a resource in designing their own courses. And as **Mickey Sarquis** and **Lynn Hogue** discuss in chapter 10, the 2nd edition of *Chemistry in the National Science Education Standards* also offers important guidance to professional development providers.

Henry Heikkenen and **Kelly Deters** begin this 2nd edition by illuminating tensions naturally inherent in the standards, e.g., means vs. ends — is one more important than the other, and how can a chemistry teacher bring them into balance? In chapter 2, **William Carroll** (former ACS president) and **Kristin Sherman** tackle many important questions. What is chemistry as practiced today? Why should we teach chemistry in high school? What is the power of a background in chemistry for today's student, given that not everyone will become a chemist? How can chemistry help manage our planet's finite resources? Their answers to these questions will help high school teachers engage their students in conversations about chemistry, careers, and the importance of chemistry in their everyday lives.

While making connections between chemistry and everyday life may be natural for chemistry teachers, it can be more difficult to identify connections between chemistry and the other sciences. In chapter 3, **Kathy Kitzmann** and **Charlotte Otto** discuss the role of unifying themes in the Standards, such as scale and structure, and how to use them for district/building planning and curriculum alignment. Chapter 5, written by **Deborah Herrington**, **Ellen Yezierski**, and **Rebecca Caldwell**, is packed with examples of how to create connections between biology and chemistry courses in the high school. In chapter 6, **Ann Benbow** and **Cheryl Mosier** make the case that chemistry as the central science can help high school students see the role of chemistry in the geosciences. They offer suggestions on how to interweave chemistry in the study of the hydrosphere, atmosphere, geosphere, and biosphere.

Leo Sorel

NSES does not mandate any particular pedagogy, which often is surprising to those who think that surely the *NSES* would recommend inquiry as a preferred pedagogy. In fact, the *Standards* make a strong case that inquiry is so important it is not just pedagogy, but rather makes the case for inquiry as *content*. Chapter 4 highlights two powerful programs to develop inquiry in high school. **Rick Moog** and **Laura Trout** describe how to use POGIL (Process-Oriented Guided Inquiry Learning) in the high school classroom while **Dawn Rickey** and **Chris Lee** make a case for MORE (Model Observe Reflect Explain) learning in the laboratory.

The chemistry laboratory has changed considerably since the first edition of this book was published. In chapter 7, **Loretta Jones** and **Seán Madden** highlight many tantalizing possibilities for integrating technology into the chemistry laboratory such as graphing calculators and probes for data collection. For teachers working with limited budgets or searching for novel ways to minimize safety hazards and waste disposal, virtual laboratories are also discussed.

Beyond the typical content standards, the *NSES* also emphasize the importance of historical and personal perspectives. **Donald Wink**, **Patrick Daubenmire**, **Sarah Brennan**, and **Stephanie Cunningham** offer their collective wisdom about incorporating these in a systematic manner, primarily with examples drawn from an implementation of the ACS textbook *ChemCom* (*Chemistry in the Community*). Chapter 8 chronicles their work in the Chicago Public Schools to connect chemistry to the daily lives of their urban students. And in a corollary to connecting chemistry to the 21st century, **Seth Rasmussen**, **Carmen Giunta**, and **Misty Tomchuk** remind us in chapter 9 that when teaching our students about the history and nature of science, we should avoid the temptation to "sanitize" the lessons for students. Rather, they discuss the importance of students learning that science progresses through "starts" and "stops."

One very big change since the first edition of this book was published in 1996 is the enactment of No Child Left Behind. Given the realities of NCLB funding and the mandate for assessment, high school chemistry teachers will be eager to read chapter 11. **Tom Holme** and **Laura Slocum** discuss how to improve the writing of items to test students' chemistry knowledge. They also describe professional development opportunities for high school chemistry teachers through the ACS Examinations Institute, and how Institute exams are aligned with the NSES. Another change under way as this book is being published is a major reform of AP courses by the College Board. In chapter 12, **Jim Spencer** and **John Hnatow** present the rationale for changing the AP chemistry curriculum as well as the "Unifying Themes" that are guiding the AP chemistry curriculum revisions.

The 2nd edition of this book features three new chapters. **Doris Kimbrough** and **Susan Cooper** offer practical solutions in chapter 13 to the chemistry teacher who has students trying to learn English and the language of chemistry at the same time. **Dorothy Gabel** and **Karen Stucky** provide a summary of the prior knowledge that high school chemistry students bring with them from their K–8 learning; chapter 14 also includes a brief review of misconceptions common among high school chemistry students. Explaining how misconceptions develop and using research on how students learn to offer practical suggestions for chemistry teachers are topics addressed by **Diane Bunce**, **Sharon Hillery**, and **Elena Pisciotta** in chapter 15. The book ends with a forward-looking, thought-provoking discussion of the challenges chemistry teachers will face in the future. In chapter 16, **Steve Long** and **Mary Kirchhoff** share how the American Chemical Society and its Education Division stand ready with resources to empower teachers to meet these new opportunities.

Of course, a 2nd edition of *Chemistry in the National Science Education Standards* would not be possible without the authors whose contributions were responsible for the first edition. Thank you to Ronald Archer, Jerry Bell, Ann Benbow, Bonnie Brunkhorst, Diane Bunce, Dwaine Eubanks, Lucy Eubanks, Henry Heikkenen, Stanley Pine, Kathryn Scantlebury, Patricia Smith, Conrad Stanitski, Michael Tinnesand, Mary Virginia Orna, and Sylvia Ware.

In the introduction to the 1st edition of this book, Stanley Pine wrote:

> *"The standards are the cornerstone of our ongoing national efforts to improve the quality of science education for all our students. Their implementation will require a long-term effort and adequate support from educators, policy makers, and the broader public in order to accomplish the stated goals."*

It bears mention that the children born in 1996 are just today in the 5th and 6th grades. These children are just halfway through learning the science they should know and be able to do. It is our hope that this 2nd edition will empower high school chemistry teachers to meet students with knowledge of what they already know, how they learn, and the central role that chemistry can and should play in their high school education. It is also our hope that, as chemistry and chemistry teaching continue to change, a 3rd edition of this book will capture the ever-evolving nature of chemistry education. We welcome your comments regarding this edition and suggestions for future volumes.

Stacey Lowery Bretz, editor
Miami University
Department of Chemistry & Biochemistry
Oxford, OH 45056

March 2008

Chapter 1

Thinking About Standards

by Kelly M. Deters and Henry W. Heikkinen

Kelly M. Deters *is a chemistry teacher at Shawnee Heights High School in Tecumseh, KS. A National Board-Certified teacher, she is a state and nationally recognized educator, high school chemistry textbook author, founder and past-president of the Kansas Association of Chemistry Teachers. Kelly has presented workshops at all levels, from building professional development to national meeting symposia presentations. She is currently pursuing her doctoral degree from the University of Nebraska at Lincoln. Contact e-mail: kellymdeters@gmail.com*

Henry W. Heikkinen *is Professor Emeritus of Chemistry and former Director of the Mathematics and Science Teaching Institute, University of Northern Colorado. He participated in the original development of the National Science Education Standards, co-chaired the Colorado State Model Content Standards working group, served as a consultant to the American Association for the Advancement of Science Project 2061, and was a former member of the College Board's Science Academic Advisory Committee. Contact e-mail: heikk2000@comcast.net*

The basic foci of chemistry teaching in schools have remained unchanged since chemistry became a secondary school course in the early 1800s. These constant concerns are encompassed by just three questions:

- What should students know, understand, and be able to do?
- How will they get there?
- How will we know if they actually attained these goals?

Although those questions have remained unchanged, the particular answers offered at a particular time depend on what is known and valued about chemistry, learning, instruction, assessment, and technology, as well as prevailing views about societal priorities and the purposes of schooling.

Two sets of comprehensive national guidelines were developed in the 1990s that provide contemporary answers to those three questions:

- *National Science Education Standards* (NSES)—produced under the aegis of the National Research Council (NRC, 1996); and
- *Benchmarks for Science Literacy*—developed by Project 2061 of the American Association for the Advancement of Science (AAAS, 1993).

These documents have influenced and energized science education reform planning, policies, and action at national, state, and school district levels since their publication over a decade ago, as well as addressed the three questions posed earlier:

- *What should students know, understand, and be able to do?* NSES content standards and Project 2061 benchmarks provide answers in terms of fundamental concepts and skills to be learned by every student.
- *How will they get there?* NSES science teaching standards and Project 2061 tools and blueprints offer criteria for effective instruction and curricula. NSES program and system standards also identify external support needed to create and sustain standards-based science instruction in classrooms.
- *How will we know whether they have reached the goals?* NSES assessment standards and Project 2061's assessment blueprints provide guidance for monitoring student progress in attaining standards-based goals and suggestions about how such assessment data can inform and guide instruction.

David Armer, USNCO

Consistent with the national standards-based agenda, we can clarify the actual goal of chemistry education today: The ultimate goal of chemistry education reform efforts is *not* to improve the quality of classroom instruction, develop better textbooks or teaching units, implement better laboratory activities, or use more authentic assessments. Nor is the goal to implement new instructional methods, encourage group work, or even to use "hands-on" experiences. While these approaches certainly possess merit, their value is as a *means* to a common, well-focused *end* or goal: *improved student learning of central facts, ideas, and skills of chemistry*.

From a standards-based perspective, the sole "quality claim" that counts for *any* instructional technique is that it should clearly contribute to students' improved science learning. Instructional strategies are all *means* to an *end*, not ends in themselves. In other words, content standards express the learning *ends*, and not the *means* by which they may be reached.

Now that learning goals in science for all K–12 students have been established nationally (with related state and local efforts completed), what do "standards" imply for improving science (and thus chemistry) teaching and learning? Educators, parents, and community leaders have developed their own understanding about implications of this standards-based agenda. Unfortunately, those well-intentioned efforts have generated some unfocused, misleading notions. Those common misconceptions about standards-based science education are sufficiently pervasive to justify their attention here, as they may influence how ideas discussed later in this book are understood and interpreted.

Standards vs. Content Standards

"Implementing standards" is sometimes narrowly understood to imply that the K–12 science curriculum faithfully supports student learning defined by *content* standards. However, NSES insists that *all five* additional science-standard categories, carefully described by the NSES document (NRC, 1996), must be addressed: *science teaching* (instructional) standards (discussed in this book in chapters 4, 13–15), *professional development* standards (discussed in chapter 10), *assessment* standards (discussed in chapter 11), K–12 *program* standards, and school *system* standards. Content standards are merely an initial direction-setting step toward comprehensive standards-based reforms. Clearly, that decade-old NSES document remains highly relevant today; it certainly should not be dismissed as "yesterday's news."

Standards-Based vs. Standardized

These sound-alike terms cause mischief and confusion, since *standards-based* vaguely resembles *standardized*. The latter term, often associated with traditional school testing, carries historic baggage of rigid, fact-based, low-level test items. By contrast, classroom assessments designed to probe student learning of material defined by content standards (*standards-based*) ensure that these important student learning goals are validly assessed.

Content Standards vs. Curricula

Content standards do not specify particular science courses or curricula; they just map ideas and skills that all students—over their K–12 schooling—should know, understand, and be able to do (AAAS, 2001). Learning *ends* (content standards) are distinct from instructional *means* used to reach them. Specifying a journey's destination doesn't imply any particular travel route or mode of transportation. NSES address both standards-based *ends* (content standards), as well as standards-based *means* (teaching standards, assessment standards, etc.).

Content standards have roughly the same relationship to a school's curriculum as nutritional standards have to a particular diet or cuisine. People don't consume nutritional standards; they consume well-prepared meals, ensuring that desired nutritional standards are met. Nutritional standards can certainly guide the design of various diets and can help evaluate their quality, but they don't dictate or define a particular sequence of meals. (How many people elect to consume their entire daily allotment of dairy products within one meal?) Likewise, strictly speaking, schools do not *teach* (or even *implement*) content standards, but rather, schools plan, implement, and deliver sequences of instruction (curricula) to help students attain those learning goals.

Diversity vs. Unity

Despite one common set of nutritional standards, there are many cuisines, most of which meet those standards. In fact, people prize variety among their dining experiences. Similarly, content standards don't mandate particular science courses or science curricula. Many pathways exist to learning goals defined by content standards (NRC, 1999*a*, *b*). Students learn more effectively when curricula are adapted to their interests and to their instructor's interests and strengths, which vary across schools, school districts, states, or the nation. Thus, curricular flexibility should be encouraged across all levels (e.g., no mandated topic sequence within any particular courses, or predefined classroom or laboratory activities).

Where's the Chemistry?

Chemistry ideas and skills appear within *all* eight NSES content standards categories (Unifying Concepts/Processes in Science, Science as Inquiry, Physical Science, Life Science, Earth/Space Science, Science and Technology, Science in Personal/Social Perspectives, and History/Nature of Science). It is quite correct that the word *chemistry* does not appear in either the table of contents or index of *Benchmarks for Science Literacy* or *National Science Education Standards* (AAAS, 1993; NRC, 1996); however, it may be reassuring to chemistry teachers that the terms *biology*, *physics*, *geology*, or *astronomy* do not appear there, either.

The absence of these familiar science-content organizers should not trigger alarm. It does not deprecate the usefulness of those collective nouns as titles or organizers of school science courses. Rather, this observation serves as an additional reminder about *means* vs. *ends*. These documents *do not* describe courses or curricula; they just map intended science facts, ideas, and skills that all students—over their K–12 studies—should know, understand, and be able to do.

However, all central facts, ideas, and skills of chemistry are mapped collectively within *eight* categories of NSES content standards. That simply acknowledges and highlights chemistry as a "central science," with its applications and implications across all branches of natural science. Thus, one should not expect chemistry topics to appear only within Physical Science content standards (discussed in this book in chapters 5 through 9).

NSES content standards, while presented within particular categories, do *not* imply how school teaching units, courses, or curricula should be organized. Confusion on this point can be clarified, once again, by distinguishing means and ends: Just as specifying a journey's destination does not imply any particular route or mode of transportation, specifying and organizing learning *ends* (content standards, benchmarks) are distinctly different from the instructional *means* (curricula, courses) designed to reach them.

Discipline-Based vs. Interdisciplinary

Consistent with the previous point, NSES content standards understandably remain silent on merits of either discipline-based or interdisciplinary/integrated approaches to curriculum design. However, content standards ensure that cross-disciplinary or integrated discussions about science-course design can occur without any threat of "watering down" or omitting valued chemistry content for all students. Well-defined content standards ensure that all intended student science learning will remain on the curriculum-design table.

Some vs. All

When chemistry teachers are asked to align their courses with NSES, they frequently respond, "I already teach that, and more." However, aligning instruction with NSES requires more than just ensuring that chemistry courses address relevant content standards. The NSES content standards apply to *all* students, not just to those currently enrolled in chemistry (even if that's a majority of students). This carries major implications for high school science course design. How will students not electing to enroll in biology, chemistry, and physics courses still meet the standards? What alternative courses or sequences will be made available to the *rest* of students? And how will these courses meet other requirements, such as state graduation requirements or Regents Board or college entrance requirements?

In brief, content standards delineate what should be *learned* by all students, not just what should be *taught* by teachers or offered in elective courses. Courses and assessments should ensure that standards-based content is *learned* by *all* students, not just by those students completing courses "covering" that content.

Can vs. Should

Upon reflecting on the notion that content standards require that *all* students should learn that content, some experienced teachers may respond, "not all students can do that!" However, secondary school content standards do not necessarily describe what students *can currently do*, but what they *should be able to do,* following a K–8 standards-based science curriculum. If NSES were fully implemented in 1996 when it was first published, the juniors of 2007 would have possessed the knowledge and skills of a comprehensive standards-based science program. So, are we there yet? Most teachers would certainly agree that we are not. Why not?

An unexpected side effect of the initial cycle of No Child Left Behind (NCLB) legislation was that while schools directly focused funding and activities to enhance student performance in mathematics and reading, many schools relegated science education to the back burner. This placed teachers and students at a science-learning deficit; NCLB-mandated science assessments will be possibly linked to school accreditation and perhaps to Annual Yearly Progress indicators. But even if schools *had* revamped their science programs in 1996, time is still required to develop curricula, gather resources, provide in-service teacher support, evaluate progress, and adjust instructional programs. What can high school teachers do to fill a science-background gap among their current students, while they await standards-based students to enter their schools?

9–12 vs. K–12

If high school teachers want to ensure that students arrive fully prepared to meet secondary-school content standards, then they should envision a *K–12* science program rather than just a 9–12 science program. High school science teachers can meld their content knowledge with K–8 teachers' pedagogical and developmental knowledge (discussed in this book in chapter 14), with guidance from content standards, to establish a sound, vertical science curriculum. Such a curriculum should be cohesive—not envisioned as discrete instructional "layers," but as a purposeful development of ideas and skills mapped over students' 13 school years (AAAS, 2001).

Chemistry-related content standards should not be "saved up" for high school chemistry classes but should be *supported and developed over students' entire school experience.* Standards-based reforms imply that high school teachers, ideally, should confer with K–8 teachers to bolster their confidence about science content and best practices, including inquiry-focused, standards-based, developmentally appropriate science for students. Secondary school teachers can encourage integrating science with other content areas (such as mathematics, reading, writing, and other modes of communicating) to help prevent K–8 teachers from abandoning science in favor of strengthening other curricular areas. Care must be taken, however, to ensure that science stands as a valued curriculum in its own right and that other studies are linked to the science curriculum rather than, for example, merely selecting science-related prose for reading instruction and concluding that students have therefore encountered enough science. Science instruction should be championed and supported by a district's science curriculum coordinator or resource person. This admittedly is an ideal expectation—school districts may not yet possess needed time or resources to implement such standards-based "vertical thinking."

Mike Ciesielski

Floors vs. Ceilings

The focus on standards being met by *all* students and remediation for students not prepared to meet secondary school content standards provokes concern that students will be held back, and the richness and opportunities of more rigorous science courses will disappear in favor of more homogeneous instruction. However, content standards define the *floor*, not the *ceiling*. Content standards do not limit possibilities for students able to achieve more, they merely specify the *minimum for all students*. A "science for all" vision does not in any way preclude or discourage additional science-learning opportunities, enrichment, and course options for students motivated and capable of pursuing them.

Inquiry vs. Content

Historically, science teachers (and those who prepare, support, and supervise them) have often debated whether classroom emphasis should be placed on students' development of inquiry skills or on their assimilation of science content. This is a false dichotomy, according to the NSES (NRC, 1996, 2000), which defines inquiry *as content,* rather than merely as an instructional strategy to address "real content." Considered as content, inquiry encompasses both student understanding *about* scientific inquiry and the skills needed to *do* inquiry (NRC, 2000).

Hands-on vs. Inquiry

Some science teachers mistakenly assume that implementing inquiry in science classrooms necessarily means that students are busily engaged in numerous activities and with laboratory work. Although hands-on work is necessary and valuable in all science courses, it does *not* ensure that students are engaged in inquiry. NSES (NRC, 2000) considers inquiry as purposeful student activity, driven by *asking questions of nature*. Modeling scientific investigations through prescribed laboratory procedures is *not* congruent with students designing and conducting their own investigations. Students cannot learn to *plan and complete their own inquiries* without first gaining needed practice. By contrast, they can also participate in classroom inquiry without engaging in chemistry laboratory work—students can seek data or gather information from literature, case studies, or the Internet, and share, evaluate, debate, and generate ideas with their classmates. Thus, inquiry learning is *not* necessarily equivalent to conventional, hands-on laboratory activities (discussed in this book in chapter 4).

Summary and Conclusion

Many instructors believed that standards-based education would answer many questions about how to best teach science to students. The standards provide a framework to answer questions concerning what students should know, how they will get there, and how we will know when they've attained those learning goals. Content standards, teaching standards, and assessment standards have set minimum expectations for all students.

However, standards-based frameworks pose additional questions for teachers, schools, and districts. Because content standards define the ends, rather than the means, schools must now decide how they will attain those desired learning goals.

Once misconceptions are dispelled concerning what standards-based education actually implies, stakeholders can begin to address more challenging issues. High school science educators should consider, among other questions: (1) Are their educational settings congruent with all standards or just content standards? (2) What particular courses and sequences should be implemented to ensure that all students have opportunities to learn science in terms of all standards? (3) What standards-based curricula should they select or design, assuming that the curriculum should be based on a K–12 development of student ideas and skills? (4) How should standards-based science curricula address the science-background gap, while awaiting the arrival of students possessing complete K–8 standards-based science knowledge and skills?

Many strategies can be devised to address these questions, depending on particular districts, schools, and teachers. This book offers suggestions and provides examples for implementing all NSES standards within the context of chemistry instruction. It's up to individual schools to weigh suggestions and examples for implementing comprehensive, well-thought-through plans for meeting science standards by *all* students.

One way to explore the challenges and tasks ahead is to reflect on what teachers and school administrators should "know, understand, and be able to do" about implications of standards, whether those standards focus on chemistry or other valued science learning. That's what this book is all about!

Recommended Readings

National Research Council (NRC). *Designing Mathematics or Science Curriculum Programs: A Guide for Using Mathematics and Science Education Standards;* National Academy Press: Washington, DC, 1999*a*. Suggestions for translating mathematics and science standards into classroom instructional materials.

NRC. *Selecting Instructional Materials: A Guide for K–12 Science.* Washington, DC: National Academy Press, 1999*b*. Tips for reviewing curricular materials in terms of their congruence with NSES.

Recommended Web Sites

NRC. *National Science Education Standards.* Washington, DC: National Academy Press,1996. http://www.nap.edu/readingroom/books/nses/ (accessed March 13, 2008).

NRC. *Inquiry and the National Science Education Standards: A Guide for Teaching and Learning.* National Academy Press: Washington, DC, 2000. http://books.nap.edu/html/inquiry_addendum/ (accessed March 13, 2008).

References

American Association for the Advancement of Science (AAAS). *Benchmarks for Science Literacy.* Oxford University Press: New York, 1993.

AAAS. *Atlas of Science Literacy.* AAAS Press: Washington, DC, 2001. A literal mapping of science content benchmarks, indicating a sequencing of topics designed for optimal student learning.

National Research Council (NRC). *National Science Education Standards (NSES).* National Academies Press: Washington, DC, 1996. http://www.nap.edu/readingroom/books/nses/ (accessed April 10, 2008).

NRC. *Designing Mathematics or Science Curriculum Programs: A Guide for Using Mathematics and Science Education Standards.* National Academies Press: Washington, DC, 1999*a*.

NRC. *Selecting Instructional Materials: A Guide for K–12 Science.* National Academies Press: Washington, DC, 1999*b*. Tips for reviewing curricular materials in terms of their congruence with NSES.

NRC. *Inquiry and the National Science Education Standards: A Guide for Teaching and Learning.* National Academies Press: Washington, DC, 2000. http://books.nap.edu/html/inquiry_addendum/ (accessed April 10, 2008).

Chapter 2

Why Chemistry?

by William F. Carroll, Jr., and Kristin M. Sherman

William (Bill) Carroll *is a vice president of Occidental Chemical Corporation in Dallas, TX, with nearly 30 years of industrial experience. He holds a B.A. in chemistry and physics from DePauw University in Greencastle, IN, an M.S. from Tulane and a Ph.D. from Indiana University. The latter two degrees are in organic chemistry. He is adjunct professor of chemistry at Indiana, where he teaches polymer chemistry. In 2005, Bill was president of the American Chemical Society. Contact e-mail: William_F._Carroll@oxy.com*

Kristin M. Sherman *has taught chemistry in Texas for 20 years, from the middle school through college. She currently teaches AP and pre-AP chemistry at McKinney Boyd High School in McKinney, TX. Kristin holds a B.S. in chemistry and an M.Ed. in educational leadership from Stephen F. Austin State University in Nacogdoches, TX. Kristin is working toward a Ph.D. in chemistry education at the University of North Texas in Denton, TX. Contact e-mail: kristinsherman1@tx.rr.com*

Isn't it obvious? Could guidance for implementation of a document called *National Science Education Standards* (NSES) be written without a discussion of chemistry and all of the other disciplines based on chemistry and chemical principles? After all, chemistry is a fundamental science. The goal of the standards is to ensure that all children and eventually all citizens are scientifically literate. Surely, chemistry must be integral to scientific literacy.

Maybe it's not so obvious. In the context of this time in history, chemistry has been identified with odors, explosions, poison, war, and cancer. The teaching and practice of laboratory work has become increasingly expensive and difficult to conduct safely and competently. Could a student become sufficiently scientifically literate by studying and understanding biology, physics, earth science, and meteorology? Perhaps chemistry might be an unaffordable luxury.

So for a moment, as chemists and teachers of chemistry, let us suspend our disbelief at that last statement and carefully consider the case for chemistry.

Chemistry fits neatly between the largely macro world of biology and the largely micro world of fundamental physics, and, in a sense, they both depend upon chemistry. Biological processes of cellular operations and organism reproduction are driven by chemistry. Exploration of physics would be impossible without man-made materials, such as advanced materials of particle accelerator construction and NASA's Gravity Probe B—and the chemistry that produces them.

In 2005, the American Chemical Society devoted the year to identification of a new vision for the Society. Thousands of member opinions were solicited and digested, which eventually led to the following: "Improving people's lives through the transforming power of chemistry." Why chemistry?

Table 1. Better Things for Better Lives

Chemistry is an academic exercise but brings its greatest value when applied to human need. Chemists create medicines that cure and manage diseases and allow longer, happier, and more productive lives. They create the materials—plastics, semiconductors, alloys, composites—that keep food fresh and safe, enable our computer-driven society, eliminate corrosion, and make vehicles stronger, lighter, and safer.

Chemical inventions range from medical necessities to everyday products that make life easier. The National Historic Chemical Landmarks program, administered by the American Chemical Society, celebrates many of these inventions.

Selman Waksman and Antibiotics. Waksman and his students, in their laboratory at Rutgers University, established the first screening protocols to detect antimicrobial agents produced by microorganisms. This deliberate search for chemotherapeutic agents contrasts with the discovery of penicillin, which came through a chance observation by Alexander Fleming. During the 1940s, Waksman and his students isolated more than 15 antibiotics, the most famous of which was streptomycin, the first effective treatment for tuberculosis.

Nylon Changes Fabric of Life. DuPont introduced nylon, the first synthetic fiber, in 1939 to compete with cotton, silk, wool, and rayon. The new product forever changed the textile industry and gave women's hosiery the name by which they are still known: nylons.

Do-it-Yourself Movement Born in Paint. The Sherwin-Williams Company developed Kem-Tone when the winds of World War II reduced the supply of petroleum, linseed oil, and other traditional paint ingredients. Company chemists were asked to create a durable paint that could be made with readily available substances, such as water. They looked to the ancient Egyptians for ideas and discovered that casein (a milk protein) mixed with varnish, water, and other ingredients produced a paint that covered in one coat and kept its color even with repeated washings.

The Columbia Dry Cell Battery. In 1896, the National Carbon Company (predecessor of Energizer) introduced the sealed, six-inch, 1.5-volt Columbia dry cell, the first battery marketed for consumer use. The technology of the Columbia, a carbon-zinc battery using an acidic electrolyte, served as the basis for all dry cell batteries for the next 60 years, until the introduction of the alkaline battery by the Eveready Battery Company (now Energizer) in the late 1950s.

The Development of Tide. Tide, the first heavy-duty synthetic detergent, debuted in 1946, the culmination of a search to replace traditional soaps, which did not clean well in hard water, where they deposited a residue of scum, or curds. Tide was not just a new product, but a new kind of product. It was based on synthetic compounds rather than natural products. Although initially targeted for marketing in areas of hard water, synthetic detergents—with Tide in the lead —soon displaced traditional soaps throughout the United States.

Of the basic sciences, chemistry is the one that most directly translates to products that people use and that can have a direct impact on their lives. Chemistry fuels an industry that reduces its inventions directly to practice. Table 1 summarizes several such transfers of innovation into practice. But most importantly, chemistry fits neatly with the case made for scientific literacy in the introduction to the NSES. To paraphrase: (1) Science literacy fosters personal fulfillment and excitement; (2) modern life requires scientific ways of thinking; and (3) scientifically engaged citizens will help society address shared responsibility and fairly manage shared resources. By substituting "chemistry" for "science," this chapter will examine these three goals in a chemistry context.

Chemistry literacy fosters personal fulfillment and excitement

"Some people will want to be chemists and find cures or invent new things."
—Lindsay, age 15 *

Personal fulfillment comes in both material and spiritual ways. Materially, well over a million people are employed by or directly dependent upon the chemistry enterprise—industry, academe, and government—in the United States. While many people think of careers in chemistry as research based, and of course, many are, there are also many careers besides research.

The chemical industry operates safely, effectively, and efficiently because of process development, oversight, and maintenance by chemical engineers. Chemists in government conduct research, but also develop and implement regulations and policy that foster continuous improvement and protection of the environment. Chemical analysis is fundamental to a number of industries and government agencies.

In fact, the 21st century is the era of the "nontraditional" career in chemistry. Lisa Balbes (2007) has described chemists who have careers in information science, patent law, sales, marketing, business development, and even in public policy.

As a profession, chemistry remains economically desirable. Chemists experience similar rates of unemployment as other holders of equivalent college postsecondary degrees, but chemists' salaries greatly exceed the average for each degree level (Table 2).

Additionally, undergraduate chemistry is a part of the curriculum leading to undergraduate degrees in most other technical professions. For example, undergraduate premedicine education requires an understanding of acid and base chemistry, the organic chemistry of pharmaceuticals, the creation and use of new polymers, and the biochemistry of human systems. Similar knowledge is also required for careers in dentistry, pharmacy, and nursing.

*Sherman polled a high school class for reasons to study chemistry. A few of the responses are reproduced in this chapter.

Table 2. Salary and Unemployment Data: Chemistry vs. All United States, 2005 (Heylin, 2006; U.S. Bureau of Labor Statistics, 2007)

Degree	Chemistry		All United States	
	Unemployment, %	Median salary, $M/year	Unemployment, %	Median salary, $K/year
Bachelor's	3.2	65.2	2.6	48.7
Master's	2.9	77.5	2.1	58.7
Doctorate	2.9	95.0	1.1	73.9

Chemistry has long been known as the central science because of its place in connecting and explaining the "how" of the other sciences. The interdependent nature of the sciences, as acknowledged in the NSES, indicates that chemistry is also critical to careers in biology, physics, geology, and agriculture, among others. Architects, engineers, and artists require an understanding of the nature of the materials they use. Cosmetologists use the chemistry of hair and makeup to obtain the best results for their clients. Firefighters, and especially fire officers, must understand the complex nature of combustion and must be ready to adapt fire suppression materials and fireground strategy accordingly. Elementary school teachers need chemistry to teach their students about the wonders of science and to answer the pesky question, "Why?"

Chemistry literacy fosters personal fulfillment and excitement. Ultimately, personal fulfillment is more spiritual and more important. Stories about high school teachers who inspired students to pursue the study of chemistry by the wonder of a chemical transformation or of chemists who dedicated their lives to the solution of a research problem solely for personal satisfaction are too numerous to ignore.

These stories include the personal stories of the authors. Sherman has shared her passion for chemistry which, in turn, ignited that same passion in some of her students. Her enthusiasm helped others not so interested in the class "to stick with it." Carroll was drawn to chemistry because of his high school chemistry experience. Most people are drawn to a career in chemistry because they love it. Chemistry satisfies their intellectual curiosity, their need to discover, to organize, to solve problems, and to understand the world around them, as can be seen in Table 1. Chemistry satisfies their need to create and to contribute to human well-being and opens the doors to other endeavors. Chemistry is fulfilling.

Modern life requires chemical ways of thinking

"You should study chemistry because it makes you smarter."
—Josh, age 16

The NSES document is clear. Standards exist to bring national consistency to education, an inherently local enterprise in the United States. The goals of the NSES are to define a "scientifically literate society," particularly where citizens "use appropriate scientific processes and principles in making personal decisions" and "engage intelligently in public discourse and debate about matters of scientific and technological concern."

Chemistry classes provide a learning platform for students to develop skills in technical writing, technical reading, data analysis, calculation, analytical thought, and working in teams—skills basic to daily life and successful employment. While not every job requires all of these skills and not every living moment requires technical analysis, our society and economy, if it is to be sustained, require people to exhibit a command of at least some of these attributes on a daily basis.

In a chemistry class, students are taught practical mathematics skills, and they learn how to find patterns in real data. Chemistry teaches the use of practical algebra skills in a setting beyond a mathematics class. Chemistry supersedes the basic algebra used to solve formulas and, like story problems, teaches students how to select the right formula for a situation.

Chemistry teaches students how to use proportions via dimensional analysis to make things larger or smaller in scale. This skill is useful in work beyond the chemistry classroom. Architects, engineers, and carpenters use proportions to make models and to design and build structures. Even in ordinary life, recipes must be doubled or cut in half depending on how many people the cook is feeding. Chemistry teaches the use of practical algebra skills that go beyond the conceptual framework of a mathematics class. Chemistry makes mathematics make sense.

Technical writing skills include logically arguing and supporting ideas, describing people and conditions, formulating cogent directions, and using clear, concise language. Technical reading skills include understanding text and extracting specific information, reading and following directions, using charts, pictures, and diagrams as sources of information, and recognizing and ignoring irrelevant material. These skills are sometimes taught in high school English classes but are heavily emphasized in chemistry.

Chemistry textbooks are designed to present information in a variety of formats. A functionally literate adult must be able to read and extract information presented in diverse forms; a chemistry textbook presents information from just such a variety of sources and provides students the opportunity to learn this skill.

Laboratories—even the "cookbook chemistry" kind—provide opportunities for students to read, follow instructions, and record observations. While central to laboratory reports, clear, concise writing, supported by data, is also critical to any persuasive argument in business or law.

Data analysis skills include reading and interpreting tables, charts, and graphs, fitting data into the "big picture," predicting future events, and generating new information. Data interpretation, in the context of chemistry, requires pattern recognition, and an understanding of support, contradiction, and anomaly. This requirement is not so far removed from analyzing data in personal medical histories, deciding which car is the best for the money, or making decisions about personal investments. In chemistry, students spend much time analyzing data and interpreting the data through simple question-and-answer strategies. The goal in the class, as in life, is to make supported generalizations based on scientific concepts and to apply those concepts to new situations.

While "physics" or any other sciences could be substituted for the word "chemistry" in the preceding paragraphs, chemistry is unique in its ability to address how issues of science and technology affect people individually and globally. These issues are important to policy, and policy ultimately impacts every citizen's life. Should corn be fermented and distilled to make fuel or reserved for food? Should nuclear power replace combustion of fossil fuels for electricity? How should garbage be recycled for the best benefit to the community? Should meat be irradiated to prevent the spread of disease? How will we deal with emerging epidemics or obsolescence of common antibiotics?

Citizens do not have to develop technical answers to these questions. Sometimes, the depth of the technical arguments is even beyond the specific expertise of scientists outside a field. However, citizens must be able to understand the basic arguments of the debate of food vs. fuel; greenhouse gases vs. nuclear waste reprocessing and storage; overall safe food handling practices; and national research and emergency response priorities. Without a basic knowledge of chemistry and the other sciences, taught in the context of its application to modern life, this debate becomes opaque or oversimplified to the level of bumper sticker slogans.

Chemically engaged citizens will help society address shared responsibility and fairly manage shared resources

"We should study chemistry because it allows us to really understand about the makeup and 'how come' of everything around us."

—Ashley, age 15

Most importantly, we study and practice chemistry to improve life in the aggregate for us all. Ten years ago, Stuart Hart (1997) revisited work of Paul Ehrlich and Barry Commoner, relating the sources of environmental burden in an equation:

$$Environmental\ Burden = f(population, affluence, technology)$$

How should society address the potential or reality of increasing environmental burden? Population cannot realistically be reduced in a socially acceptable way in the short term; in fact, most demographers believe that population will increase by about 50% before leveling off in the mid-21st century (Lutz et al., 2001).

Reduction in the overall level of affluence is unacceptable, at least politically, in the more affluent countries, and may have environmental consequences: poorer economies tend to be more environmentally destructive.

In short, Hart argues, humankind will not save its way into a high standard of living and a global economy that the planet can support indefinitely. The only answer that acknowledges and possibly accommodates growing population and growing affluence is technology. But the bar is set high: easy calculation suggests that we must extract a factor of 4 to a factor of 20 times more benefit from the dwindling or more costly resources we have in exchange for the same environmental burden. That need for technology advance largely falls on chemistry.

David Armer, USNCO

Few issues loom larger for the next half-century than energy. The late Nobel Laureate Rick Smalley outlined the case in a famous lecture, "Be a Scientist—Save the World" (Smalley, 2007). He argued for the critical need for new modes of energy generation and the pivotal role that nanotechnology will play in energy efficiency. In a similar presentation, Nathan Lewis (2007) of Caltech outlines the grand challenges for science and technology in this context, including "disruptive" solar technologies, more efficient electrochemistry, conversion of CO_2 to methanol and other liquid fuels, and other storage technologies. Chemistry is critical to each of these potential solutions.

Chemists design processes through principles of Green Chemistry and Engineering that reduce resource use and impact on the environment while still fostering economic growth. Many pharmaceutical chemical processes are characterized by waste intensity; the ratio of waste generation to production is 25 to 100 (Cue, 2005). Chemists use the principles of green chemistry to devise new, more efficient processes that decrease cost and reduce production of waste.

The road to cleaner, more abundant energy, better pharmaceuticals, lower-waste processes, and advanced materials—the kind of technology required to impact Environmental Burden—goes through chemistry. Bringing chemistry to bear on environmental burden is the only practical way to approach sustainability.

Why Chemistry?

"Every person should study chemistry, at least briefly."
—Steven, age 15

There is a consensus in the United States today that education is critical for economic prosperity. The NSES document asserts that "Science is for all students." Why Chemistry? If we wish to invest our students with grounding in science and an understanding of the way the world works in the hope that it will make them better citizens of the 21st century, grounding in chemistry and its impact on our lives are a critical part of that understanding.

Recommended Readings

Cobb, C.; Fetterolf, M. L. *The Joy of Chemistry: The Amazing Science of Familiar Things.* Promethius Books: Amherst, NY, 2005.

Emsley, J. *Molecules at an Exhibition: The Science of Everyday Life.* Oxford University Press: New York, 1998.

Sacks, O. *Uncle Tungsten: Memories of a Chemical Boyhood.* Random House: New York, 2001.

Schwarz, J. H. *Radar, Hula Hoops, and Playful Pigs: 67 Digestible Commentaries on the Fascinating Chemistry of Everyday Life.* ECW Press: Toronto, Canada. 2001.

Recommended Web Sites

Brain, M. How Stuff Works. http://howstuffworks.com (accessed March 24, 2008).

Davies, S. Suite 101: Chemistry. http://chemistry.suite101.com (accessed March 24, 2008).

References

Balbes, L. *Non-Traditional Careers for Chemists.* Oxford University Press: New York, 2007.

Cue, B. W., Jr. A New Game Plan. *Chem. Eng. News* 2005, *83*, 46–47.

Hart, S. L. Beyond Greening: Strategies for a Sustainable World. *Harvard Bus Rev* 1997, January/February, 66–76.

Heylin, M. Employment and Salary Survey. *Chem. Eng. News* 2006, *84*, 42–51.

Lewis, N. Global Energy Perspective (Presentation), 2007. http://nsl.caltech.edu/energy.html (accessed April 10, 2008).

Lutz, W; Sanderson, W.; Scherbov, S. The End of World Population Growth. *Nature* 2001, *412*, 543–545.

Richard E. Smalley Institute for Nanoscale Science and Technology. Be a Scientist; Save the World. http://cnst.rice.edu/whatwedo.cfm?doc_id=1220 (accessed April 10, 2008).

U.S. Bureau of Labor Statistics. Education Pays. http://www.bls.gov/emp/emptab7.htm (accessed April 10, 2008).

Chapter 3

Chemistry and Unifying Themes of Science

by Kathy Kitzmann and Charlotte A. Otto

Kathy Kitzmann is the science department chair and a teacher of chemistry at Mercy High School in Farmington Hills, MI. Kathy earned a B.S. in chemistry from Taylor University and an M.S. in medicinal chemistry from the University of Michigan. As an experienced chemistry teacher for 33 years, Kathy has served on the Michigan Science Teachers' Association Board and chaired their conference in 2002. In 1997, she was recognized as the American Chemical Society Central Region High School Teacher of the Year. Contact e-mail: KathyK@sefmd.org

Charlotte A. Otto is a professor of chemistry and professor of science education at the University of Michigan-Dearborn. After a career in applied organic chemistry research, she started a new career in K–8 science education. She is part of a team of scientists and science educators who developed a required capstone course for all preservice K–8 teachers that combines an action research project in a K–8 school with an in-depth exploration of one of the unifying themes in science. Contact e-mail: cotto@umich.edu

In the 1990s, the American Association for the Advancement of Science (AAAS) articulated "common themes" or core concepts in both *Science for All Americans* (1990) and its companion, *Benchmarks for Science Literacy* (AAAS, 1993), followed by the unifying themes of the *National Science Education Standards* in 1996. Both the National Research Council (NRC) and AAAS define these overarching themes as concepts that appear in many, if not all, scientific disciplines. Such themes can build connections between chemistry and other sciences, as well as between chemistry and its applications in our everyday life. Knowledge of these themes may provide a framework for unifying science and a mechanism for both learning and teaching science, shifting away from thinking of science teaching and learning as occurring only in the discrete units of physics, chemistry, biology, and earth science (Table 1). Perhaps of equal importance, these unifying concepts can and should be incorporated into science learning and teaching beginning in kindergarten and continuing throughout an entire educational career. The intent of this chapter is to briefly introduce these unifying themes and to provide examples showing (1) how these themes appear in chemistry and (2) how these themes connect to other sciences.

David Armer, USNCO

Table 1. Changing Emphases

Less emphasis on	More emphasis on
Courses with little connection to other disciplines	Courses that incorporate connections to other sciences
Fragmented instruction that moves from topic to topic without connections	Integrated instruction that focuses on fundamental concepts and processes
Concepts presented in isolation from real-world applications	Concepts and processes introduced with a real-world context and explored in real-world applications
No coordination among all science disciplines to reinforce unifying themes	Coordination throughout all grades and all sciences in terms of introduction and use of unifying themes

The NRC and AAAS sets of unifying themes have many similarities (see Table 2), but a more inclusive list that combines both sets would be better. Table 2 is a compilation developed by the authors to illustrate how these themes relate to each other.

Table 2. Comparison of NSES and AAAS common Themes

NSES Unifying Theme	AAAS Common Theme
Systems, order, and organization	Systems
Evidence, models, and explanation	Models
Constancy, change, and measurement	Constancy and change
Evolution and equilibrium	
Form and function	
Energy*	Energy*

The number of unifying themes has been a topic of discussion in the past few years. For example, the NRC (2007) suggested in *Taking Science to School* that K–8 teachers should focus on a few core ideas that are expanded each year. In September 2006, the National Science Teachers Association (NSTA) asked its members about the breadth and depth of coverage of science. It found that 78% of respondents agreed and that 85% said that state standards should be centered on a few core ideas (NSTA, 2007).

The use of themes as an organizing principle in teaching science content knowledge is one way to employ unifying themes. In science, the use of unifying themes or big ideas is not only a way to organize student learning, but it is also a way of approaching the design of a course. Wiggins and McTighe (2005) describe a method of course design that engages students in inquiry and promotes learning by providing a conceptual framework for students (p. 4). Their backward design process starts by identifying desired student learning, determining acceptable data for verification of student learning, and only then moving toward planned instruction. They describe the use of "big ideas and core tasks" as a central component of desired student learning.

This chapter offers examples of how unifying themes appear in chemistry, as well as examples of how these themes might appear in biology or physics. Brief descriptions of each theme are also provided.

Systems, Order, and Organization

Systems are defined by the NSES as an "organized group of related objects or components that form a whole" (NRC, 1995, p. 116). Important aspects of a system are the presence of boundaries, related objects, flow of a resource such as energy into and out of the system and the existence of feedback. Order is the "behavior of units of matter" (NRC, 1995, p. 117) that can be predicted and described, while organization refers to a way of structuring information that helps to reveal relationships. We should comment on the NSES use of "organization." In examples provided by NSES (p. 117), the use of "organization" appears to refer to organization of information. For example, the periodic table is an organization of knowledge of elements and their reactivity. Yet, in subsequent examples, NSES appears to consider organization within nature rather than of knowledge. In this second set of examples, the NSES refers to different levels of cellular organization [cells, tissues, organs, organisms (p. 117)] or physical systems (fundamental particles, atoms, molecules). These two definitions of "organization" are preserved in the examples provided in Table 3. When teaching classification systems of organisms in a high school biology class, the biology teacher could take a few moments to introduce students to another classification scheme common in another science, e.g., chemistry and the periodic table.

Table 3. Systems, Order, and Organization

Chemistry	**Systems:** atoms, compounds, chemical reactions **Order:** classification of matter as solid, liquid, gas, or plasma **Order:** classification of types of compounds **Organization:** periodic table and periodic trends (organization of knowledge) **Organization:** electrons, neutrons, and protons as parts of atoms (organization of nature)
Biology	**Systems:** within organisms (digestive, endocrine, reproductive, circulatory, etc.) **Order:** classification of organisms (organization of knowledge) **Organization:** levels of organization with living systems (cells, tissues, organs, organisms, populations, and communities) **Organization:** food chain or food web (organization of nature)
Earth Science	**Systems:** rock systems (igneous, metamorphic, sedimentary) **Order:** layers of the atmosphere **Order:** internal structure of the earth **Organization:** topographic maps of Earth's surface (organization of knowledge) **Organization:** patterns of airflow (organization of nature)
Physics	**Systems:** machines **Order:** collective phenomena such as superconductivity and magnetism; classifications of subatomic particles **Organization:** charts showing radioactive decay pathways (organization of knowledge) **Organization:** solar systems, galaxies, superclusters (organization of nature)

Evidence, Models, and Explanation

Evidence is simply the data that scientists or students gather to answer questions. Models differ from laws or theories in that they change with the discovery of new evidence that needs to be included in an explanation. Models can be theoretical or mathematical or concrete or functional. Explanations of scientific phenomena, to be useful, must incorporate available evidence and allow for predictions of future events. A geologist could describe how seismographs provide evidence of the motion of Earth's plates and that other sciences use different motion detectors to study phenomena. For example, chemists use infrared spectrometry to study the motion of atoms in molecules. In both cases, scientists are using evidence to study and explain phenomena.

Table 4. Evidence, Models, and Explanation

Chemistry	**Evidence:** discovery of radioactivity and types of radioactivity **Evidence:** mass spectra, X-ray diffraction patterns **Evidence:** experiments leading to the discoveries of the electron, proton, neutron, and nucleus **Models:** atomic models (from billiard ball model through quantum mechanical model) **Models:** bonding models (VSEPR, hybridization) **Models:** for reaction rate (collision theory, transition-state model) **Explanation:** explanations of the direction of spontaneous reactions in terms of energy and entropy, explanation of periodic trends in terms of effective nuclear charge
Biology	**Evidence:** observations of cells or microorganisms using microscopes **Models:** models of DNA structure **Explanation:** how specific changes in organisms confer advantages for survival, how are patterns of inherited traits explained in terms of random transmission of genes.
Earth Science	**Evidence:** vibration of the earth as recorded by seismographs **Models:** convection in interior **Explanation:** plate tectonics
Physics	**Evidence:** observational data of planetary motion **Models:** description of gravitational fields **Models:** description of angular displacement **Explanation:** Newton's laws of motion

Constancy, Change, and Measurement

The behavior of objects is sometimes described in terms of how the behavior changes with time. In other instances, the object's behavior appears to be constant. For example, the energy in a system is constant, but the form of the energy may change over time. Measurement is a way to quantify change or constancy of a system or object. The tools of measurement vary with the system studied; each science has its own particular modes of measurement. A physics teacher might include descriptions of conservation of energy in terms of chemical reactions, as well as in systems.

Table 5. Constancy, Change, and Measurement

Chemistry	**Constancy:** charge of an electron, gas law constant **Constancy:** Laws of Conservation of Matter, Conservation of Energy, definite and multiple proportions **Change:** development of Atomic Theory is an example of change **Change:** changes in chemical composition or properties resulting from chemical reaction **Measurement:** Energy of reactions in bomb calorimeters
Biology	**Constancy:** homeostasis of temperature in warm-blooded animals **Change:** biochemical cycles (oxygen, carbon, nitrogen, sulfur, phosphorus, Krebs cycle) **Measurement:** bacterial growth rates
Earth Science	**Constancy:** the timescale of geologic change is so long that many quantities are nearly constant over human lifetimes. **Change:** rock cycle or water cycle **Change:** soil erosion, weather patterns, and mechanical weathering **Change:** riverbed changes with erosion **Measurement:** rates of motion of continental plates
Physics	**Constancy:** the speed of light; the total mass and energy in the universe **Change:** acceleration **Measurement:** in terms of the metric system **Measurement:** using coordinate systems

Evolution and Equilibrium

Evolution is defined by the NSES as a "series of changes, some gradual and some sporadic" that result in the behavior of organisms or objects that can be seen today. Although many associate evolution with diversity of biological organisms, it also applies to nonliving objects such as the solar system. Equilibrium, sometimes grouped with constancy and change, occurs when two processes or forces are in balanced opposition to each other. Chemical reactions reach a state of equilibrium when the rate of the forward, product-producing reaction is

Table 6. Evolution and Equilibrium

Chemistry	**Evolution:** chemical origins of life (amino acids → proteins → RNA → simple life forms) **Evolution:** nuclear fusion: creating new elements **Equilibrium:** chemical equilibrium, phase equilibrium, solubility equilibrium, acid-base equilibrium
Biology	**Evolution:** changes in DNA leading to diversity of species **Evolution:** changes of an organism to adapt to its environment **Equilibrium:** maintenance of blood pH by removal of CO_2
Earth Science	**Evolution:** formation of the planets **Evolution:** changing continents **Equilibrium:** various cycles (see above)
Physics	**Evolution:** formation of galaxies, solar systems **Evolution:** thermal equilibrium **Equilibrium:** rotational and translational equilibrium

balanced by the rate of the reverse, reactant-producing reaction. Biologists, when introducing students to species evolution, could set organismal evolution in a broader context of how other objects such as stars or chemical elements have changed over time.

Form and Function

Form and function are related to each other. The form of an object informs the observer about the function, and the function provides information about the form. For example, a plant cell has rigid cell walls that relate to the overall structure of the plant. Students should be able to infer the strength of the cell wall by noting the structure of a plant. A chemistry teacher can build upon this knowledge by leading students to understand how atomic bonding orbitals provide information about molecular shape that, in turn, ultimately relate to the rigidity of plant cell walls.

Table 7. Form and Function

Chemistry	**Form:** electron configurations of elements related to ability to gain, lose, or share electrons **Form:** molecular shape determines polarity, which determines intermolecular forces, which determine function **Form:** the unique properties of water relate to its functions **Form:** carbon's form correlates to its many possible molecules **Function:** characteristic reactions of organic molecules relate to the structure (or form) of functional groups
Biology	**Form:** differences between plant and animal cells relate to function **Form:** molecular structure of DNA and RNA influences functions **Function:** organs are functioning groups of specialized cells
Earth Science	**Form:** the different mineral forms of elements found in the earth and how they are recovered **Function:** convection currents in earth's interior determine behavior of continents
Physics	**Form:** shape of lens and whether it causes convergence or divergence of light **Form:** surface structure, whether rough or smooth, determines friction and how objects move (function) on the surface **Function:** the mechanical advantage of a simple machine is related to its form

Energy

All sciences are concerned with sources and transformations of energy. Physicists view energy mostly in terms of conservation of energy *within* a system, while biologists tend to think of energy flow *through* a system. It can be argued that chemists include both views, as the first law of thermodynamics accounts for energy in terms of conservation, but when chemists describe reactions as endothermic or exothermic, they are employing an analogy of energy flows into or out of the chemical reaction system. Biology and physics teachers alike could describe these different perceptions of energy in their courses.

Table 8. Energy

Chemistry	Conservation of energy within defined systems
	Spontaneity of reactions and the "drives" to minimum energy and maximum entropy; Gibbs free energy
	Bond formation and energy changes
	Average kinetic energy of gas molecules
	Activation energy and reaction rates
	Ionization energy and electron affinity
	Thermochemistry: principles of heat flow
	Energy levels in atoms: ground state vs. excited states
	Electrochemistry (voltaic cells vs. electrolytic cells)
Biology	Cycle of energy and matter through living and nonliving systems
	Biological effects of radiation
Earth Science	Changes in the earth's ecosystem powered primarily by the energy of the sun
	Renewable energy resources
Physics	Types of energy: mechanical, electrical, nuclear, heat energy, radiant energy, chemical energy
	Work and energy (laws of thermodynamics)
	Nuclear energy; nuclear binding energy; fission and fusion
	Mass-energy relationships

Conclusion

Ideally, students would construct their knowledge of these themes beginning in kindergarten and deepen their knowledge as they progress through high school and into college. The typical disciplinary focus experienced by high school science teachers in their educational experience tends to prevent deep understanding of these themes as they appear in other sciences. The lack of focus on these same themes in general science content courses of the typical curricula of elementary and middle school science teachers similarly prevents them from gaining a deep knowledge of these themes.

As teachers of chemistry, we cannot address the lack of attention to these commonalities in other science disciplines. We can, however, incorporate these common themes in our own classroom instruction in chemistry. We can help our students see how these themes provide connections between and distinctions among chemistry and other sciences such as biology, earth science, and physics. Building these connections necessitates the use of unifying themes when we teach and learn chemistry but also requires describing how those themes have or will appear in other courses for our students.

References

American Association for the Advancement of Science (AAAS). *Science for All Americans*. Oxford University Press: New York, 1990.

AAAS. *Benchmarks for Science Literacy*. New York: Oxford University Press, 1993.

National Research Council (NRC). *National Science Education Standards*. National Academies Press: Washington, DC, 1995.

NRC. 2007. *Taking Science to School: Learning and Teaching Science in Grades K–8*. National Academies Press: Washington, DC.

National Science Teachers Association (NSTA). *Getting to the Core of Science Standards*. NSTA Reports, 2007. http://www3.nsta.org/main/news/stories/nsta_story.php?news_story_ID=53706. (accessed March 14, 2008).

Wiggins, G.; McTighe, J. *Understanding by Design*, 2nd ed. Pearson Merrill Prentice Hall: Upper Saddle River, NJ, 2005.

Chapter 4

Inquiry Learning: What Is It? How Do You Do It?

by Laura Trout, Chris Lee, Rick Moog, and Dawn Rickey

Laura Trout *has a B.S. in chemistry from Central Washington University and an M.S. in chemistry from the University of Washington. She has taught high school chemistry for 14 years, currently teaching chemistry I at Lancaster Country Day School in Lancaster, PA. Laura has used and written Process-Oriented Guided Inquiry Learning activities for her classes for about 6 years. Over the past 4 years, she has served as a POGIL workshop facilitator. Contact e-mail: troutl@lancastercountryday.org*

Chris Lee *has been teaching chemistry for 11 years and also serves as Student Data Analyst, Science Department Chair, and instructional coach for Fort Collins High School, in Fort Collins, CO. He has been involved in advancing inquiry in chemistry education through part-time research with the Colorado State University, Department of Chemistry, and now serves on the Board of Advisors for the Hach Scientific Foundation. Contact e-mail: jolee@psdschools.org*

Rick Moog *is professor of chemistry at Franklin and Marshall College and is the project coordinator for the National Science Foundation-funded Process-Oriented Guided Inquiry Learning (POGIL) Project (DUE-0231120, 0618746, 0618758, 0618800). He has coauthored POGIL materials for college-level general chemistry and physical chemistry and is coeditor of the ACS Symposium Series Book Process-Oriented Guided Inquiry Learning. Contact e-mail: rick.moog@fandm.edu*

Dawn Rickey *is an associate professor of chemistry at Colorado State University, where her research focuses on the relationships among metacognition (monitoring and regulation of one's own thinking), conceptual change, and the ability to solve novel (transfer) problems in chemistry. Dawn codeveloped the Model-Observe-Reflect-Explain (MORE) Thinking Frame as part of her graduate work at University of California, Berkeley. Contact e-mail: rickey@lamar. colostate.edu*

Over the past century, there has been an increasing accumulation of evidence on the effectiveness of inquiry approaches to science instruction at all levels. In this chapter, the principles of inquiry are presented as they relate to the *National Science Education Standards* (NSES) (NRC, 1996), along with a discussion of the research results that support this approach. This is followed by a description of how inquiry-based learning experiences differ from traditional instruction and some suggestions for how to implement inquiry approaches successfully in the high school science classroom. The chapter ends with a more detailed description of two current, research-based examples of effective inquiry pedagogies, Process-Oriented Guided Inquiry Learning (POGIL) and the Model-Observe-Reflect-Explain (MORE) Thinking Frame.

What Is Inquiry Learning?

Inquiry refers to the evidence-based process that scientists engage in to study and propose explanations about aspects of the natural world. When applied to students in science classrooms, *inquiry learning* generally indicates student participation in activities and thinking processes similar to those employed by scientists. The NSES (NRC, 1996) emphasize three important and interrelated learning goals for all students studying science: learning about the nature of science and the work that scientists do; learning to do science (i.e., developing the abilities to design and conduct scientific investigations); and understanding scientific concepts and principles. Since all three of these aspects of science learning can be facilitated by engaging students in inquiry learning, the NRC considers inquiry to be both science content and an exemplary method of teaching and learning science.

Specifically, for students in grades 9–12, the NSES indicate that the fundamental cognitive abilities necessary for students to do scientific inquiry are

- identifying questions and concepts that guide scientific investigations;
- designing and conducting scientific investigations;
- using technology and mathematics to improve investigations and communications;
- formulating and revising scientific explanations and models using logic and evidence;
- recognizing and analyzing alternative explanations and models; and
- communicating and defending a scientific argument.

Developing these abilities requires students to integrate skills such as observation and inference with content knowledge, scientific reasoning, and critical and reflective thinking to enhance their understanding of science.

In addition, according to the NSES, the fundamental understandings about scientific inquiry that students should develop during grades 9–12 are

- Scientists usually inquire about how physical, living, or designed systems function.
- Scientists conduct investigations for a wide variety of reasons.
- Scientists rely on technology to enhance the gathering and manipulation of data.
- Mathematics is essential in scientific inquiry.
- Scientific explanations must adhere to certain criteria; for example, a proposed explanation must be logically consistent, it must abide by the rules of evidence, it must be open to questions and possible modification, and it must be based on historical and current scientific knowledge.
- Results of scientific inquiry—new knowledge and methods—emerge from different types of investigations and public communication among scientists.

In part because of the breadth of the science teaching standards recommended in the original NSES document, an addendum entitled *Inquiry and the National Science Education Standards* (NRC, 2000) distilled this information into five essential features of inquiry that must be integrated into science teaching at all levels to meet the standards. Science teaching and learning sequences that meet the NSES engage students in

- investigating scientifically oriented questions;
- establishing criteria for evidence;
- proposing explanations;
- evaluating explanations; and
- communicating explanations.

Although it is important for students to generate *some* of the scientific questions that they investigate during each of their secondary science courses, they need not always (or even most of the time) generate these questions. The key is for students to be engaged in the five processes listed above.

Forms of Inquiry Instruction and Their Effectiveness. Although inquiry-based instructional methods are defined by engaging students in the construction and evaluation of scientific explanations based on evidence, it is important to note that a wide variety of instructional methods are labeled as "inquiry" by instructors and science education researchers, and that all are not equally effective for promoting student understanding. For example, instructional methods termed "open inquiry" usually involve students designing their own experiments to address some general topic, while those labeled "guided inquiry" or "discovery" usually involve students looking for patterns in data collected via given experimental procedures. Unfortunately, such terms are not always used consistently, so it is important for teachers to work to understand what a particular instructional method entails, ensuring that it incorporates the five key aspects of inquiry emphasized by the NSES, before making the decision to adopt it for their science classes.

In addition, along the continuum of instructional philosophies from teacher-controlled, didactic teaching (found in traditional lectures and "cookbook" laboratory experiments, for example) to student-controlled discovery learning, guided approaches have been shown to maximize the likelihood that students will reflect upon relevant concepts and engage in processes that promote better understanding (Hofstein, 2004; Hofstein and Lunetta, 1982; Lazarowitz and Tamir, 1994; Rund et al., 1989). Studies of students' understanding of science ideas after instruction provide clear evidence that traditional, didactic teaching methods are not very successful in bringing about productive changes in students' conceptions (Bodner, 1991; Cros et al., 1986, 1988; Gabel et al., 1987; Gunstone and White, 1981; Nakhleh, 1992; Smith and Metz, 1996). Although didactic styles of instruction can be reasonably successful in imparting the facts, rules, procedures, and algorithms of a domain, they are not effective for helping students refine and build on their ideas about science concepts, in part, because they neither require nor encourage

Mike Ciesielski

high levels of metacognition (thinking about their own thinking) on the part of the students (Rickey and Stacy, 2000). Typically, students are simply told the "correct" scientific ideas and are expected to understand them, despite the fact that they are given few opportunities and little guidance to develop such an understanding.

At the opposite end of the spectrum are "pure" discovery-learning approaches to instruction. Proponents of pure discovery believe that students should be encouraged to explore their environments creatively and that these explorations should not be curriculum driven, but based on the interests of the students (Papert, 1980). However, as with didactic approaches, discovery learning methods also fail to encourage student reflection. In fact, unguided discovery-learning methods rely on the assumption that students already possess advanced metacognitive abilities (White, 1992; Vye et al., 1998). Students in highly unstructured environments are never forced to confront their misconceptions nor are they given the opportunity to reconcile them with scientific conceptions. In addition, pure discovery methods lack sufficient guidance, and students may end up confused, not knowing what to do for long periods of time. In fact, a high degree of open-endedness in chemistry laboratory classes has been found to be significantly *negatively* correlated with achievement on chemistry examinations (Riah and Fraser, 1998).

As discussed in more detail below, the goal of a guided learning environment is to strike an appropriate balance between didactic teaching and discovery learning, allowing students to take a large measure of responsibility for their own learning, but also requiring them to reflect upon and explain their ideas, and to justify their use of evidence as well as their conclusions. Students should ultimately be challenged to think about what to do and how to do it, but given enough instructional support along the way so that they do not flounder. The challenge is to develop curricula and instructional methods such that the optimal amount of support is provided for each student. Both the Process-Oriented Guided Inquiry (POGIL) and Model-Observe-Reflect-Explain (MORE) inquiry methods, discussed in detail later in this chapter, are designed to support guided discovery in chemistry learning.

How Inquiry Instruction Differs From Traditional Chemistry Instruction.

Although inquiry-based chemistry courses have taken varied forms, all depart from the traditional teaching method in which there is a presentation of scientific principles followed by experiments to verify those principles. Traditional high school chemistry courses are typically broken down into separate lecture and laboratory components. In lecture mode, the instructor usually presents concepts to be learned, while the students listen and take notes. Students may also ask a few questions, but the participation of students in constructing scientific explanations is usually minimal.

After a concept has been presented in class, it is typically reinforced through a laboratory exercise. Traditionally, students perform a "cookbook"-style procedure that has been selected by the instructor, record data into their notebooks (or into empty spaces on a report form), and calculate values that confirm what the instructor has previously presented in the lecture. This method of laboratory instruction does not provide students with opportunities to engage in the key inquiry activities of proposing, evaluating, and communicating explanations for the chemical phenomena they investigate in the laboratory. It also does little to deepen students' understanding of the phenomena under study (Hofstein, 2004; Hofstein and Lunetta, 1982). In contrast (as will be illustrated with examples using POGIL and MORE), inquiry learning engages students in *constructing* evidence-based explanations, as opposed to simply *receiving* or *confirming* scientists' explanations of chemical phenomena.

In an inquiry-based classroom, because of the emphasis on students developing explanations based on evidence, the lecture and laboratory components of a high school chemistry course can become difficult to distinguish one from another. Scientific investigations, driven in part by student ideas, are incorporated into the "lecture" component of class. Whole-class discussions, focused on making sense of experimental observations in terms of what is happening on the molecular level, are commonplace during the "laboratory" component of the class. For example, an introduction to a new topic could begin with the instructor proposing an experiment and asking students to predict what they think will happen. After the instructor performs an experiment as a demonstration, the students would be encouraged to reflect upon what they observed, evaluate their predictions in light of the experimental evidence, and discuss what changes they might want to make to their molecular-level explanations to be consistent with the results of the demonstration. This contrasts with the traditional approach, in which the teacher carries out a demonstration and explains the results to students without involving them in the process of proposing, evaluating, and refining their own scientific explanations based on evidence.

Process-Oriented Guided Inquiry Learning

Process-Oriented Guided Inquiry Learning, or POGIL, is an instructional paradigm based on many of the research-based principles of effective instruction described previously. A POGIL classroom or laboratory experience is characterized by several common components:

- students work in small groups (usually of 3 or 4) and they generally have assigned roles;
- the instructor's role is that of a facilitator, rather than a lecturer;
- the students work on activities that have been specifically and carefully designed, usually

based on the Learning Cycle Approach (Abraham, 1998, 2005; Lawson, 1995; Lawson et al., 1989); the activities are not just "hard problems from the end of the chapter" that the students work on together; and

- the students reflect on their learning and the learning process.

Thus, the goal of POGIL is not only to develop content mastery through student construction of understanding, but also to enhance important learning skills such as critical thinking, problem solving, and assessment.

POGIL activities are typically structured to follow the three phases of the Learning Cycle. In the first phase, "Exploration," students seek a pattern in information presented to (or obtained by) them. A series of carefully designed questions leads the students to make sense of this information and to identify any inherent patterns or trends. In the second phase, "Concept Invention" or "Term Introduction," the guiding questions lead students to develop a concept from the information, and a new term can be introduced to describe this concept. In this way, new terms are introduced *after* the learner has developed a mental construct to which the term is attached. (This contrasts with the typical presentation in a textbook or lecture, in which the introduction of new words commonly occurs first, followed by examples of their use.) Finally, in the "Application" phase, students are required to have an understanding of the concept by applying it in new situations, often requiring the use of deductive reasoning skills (Abraham and Renner, 1986; Lawson, 1999). Thus, this structure guides students to construct their own understandings of a concept, imparting not only a sense of ownership in the process, but also providing the student with insight into the nature of scientific inquiry.

An example POGIL activity designed to introduce the components of an atom will clarify these ideas (Moog et al., 2006). Typically, a lecturer would tell students that atoms are composed of protons, neutrons, and electrons, and that the number of protons in the atom is known as the "atomic number" and determines the atom's identity. A POGIL activity dealing with these ideas is very different. The activity (see Figure 1) begins with a series of diagrams providing examples of a number of atoms, identifying the corresponding element and the number and location of the protons, neutrons, and electrons in each. Through a series of guiding questions, the students are led to recognize that all of the atoms with the same number of protons are identified as the same element (for example, six protons in the case of carbon). They would also note the correspondence of this number (6) with the number on the periodic table that identifies carbon. Only at this point, after the concept has been developed, would the term "atomic number" be used to describe the number of protons in one atom of a given element. In this way, an "exploration" of the information presented in the diagrams allows each student to develop the

Figure 1. An example POGIL activity. [adapted from R. S. Moog and J. J. Farrell, Chemistry: A Guided Inquiry, 3rd Edition. 2006. John Wiley & Sons: Hoboken, NJ. [Used with permission]

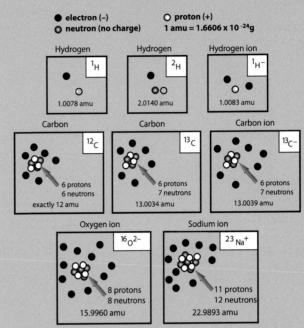

The **nucleus** of an atom contains the protons and the neutrons. amu, atomic mass units.
^{1}H and ^{2}H are **isotopes** of hydrogen. ^{12}C and ^{13}C are **isotopes** of carbon.

Critical Thinking Questions

1. How many protons are found in ^{12}C? ^{13}C? ^{13}C^{-}?
2. How many neutrons are found in ^{12}C? ^{13}C? ^{13}C^{-}?
3. How many electrons are found in ^{12}C? ^{13}C? ^{13}C^{-}?
4. a) What feature distinguishes a neutral atom from an ion?
   b) Provide an expression for calculating the charge on an ion.
5. On the basis of the model,
   a) what do all carbon atoms (and ions) have in common?
   b) what do all hydrogen atoms (and ions) have in common?
   c) How many protons, neutrons, and electrons are there in one atom of ^{1}H^{+}?
6. The number above each atomic symbol in the periodic table is called the atomic number. What is the significance of the atomic number?
7. On the basis of your answer to CTQ 6, what do all nickel (Ni) atoms have in common?
8. What structural feature is different in isotopes of a particular element?
9. The mass number, A, is the left-hand superscript next to each atomic symbol, as shown in the model. How is the mass number determined (from the structure of the atom)?
10. Where is most of the mass of an atom, within the nucleus or outside of the nucleus? Explain your reasoning using grammatically correct English sentences.

Figure 2. An activity to construct representations of some atoms and ions.

Building Atom Models

In class today, you will build models of several atoms. These models will be used in later classes to explore how the number of protons, neutrons, and electrons in an atom affect the atom's identity and properties. As you build your models, you may notice patterns in the numbers, but it is not expected that you fully understand why these patterns exist or what the consequences of them are.

In the materials packet provided, you should find:

3 permanent markers	40 red beads	43 metal beads
8 small zip-top baggies	41 blue beads	

1. The chart below lists the number of protons, neutrons, and electrons in several atoms. Divide the work evenly among group members so that each person is building only a few atoms. You need one complete set of models for your group when you are finished.

	Symbol	Atomic Mass (amu)	No. of Protons	No. of Neutrons	No. of Electrons
Hydrogen atom(a)	^{1}H	1.0078	1	0	1
Hydrogen atom(b)	^{2}H	2.0140	1	1	1
Hydrogen ion	^{1}H$^-$	1.0083	1	0	2
Carbon atom(a)	^{12}C	12.0000	6	6	6
Carbon atom(b)	^{13}C	13.0034	6	7	6
Carbon ion	^{13}C$^-$	12.0000	6	7	7
Oxygen ion	^{16}O^{2-}	15.9960	8	8	10
Sodium ion	^{23}Na$^+$	22.9893	11	12	10

2. Using a permanent marker, label each baggie with an atom's Symbol and Atomic Mass (in amus).

3. Add the appropriate number of items to the baggie to represent the atom's structure. Be sure to count carefully, as these models will be used for activities later.

Red Beads = Protons   Blue Beads = Neutrons   Metal Beads = Electrons

4. Examine the set of models and discuss any patterns you see with group members. Record your findings here.

5. Scientific models always have limitations. In what ways are the models you built a good representation of atomic structure? In what ways are the models you built a poor representation of atomic structure? (Consider the number, relative sizes, location, and charges of the subatomic particles.)

6. How could these models be improved to better represent the actual structure of atoms?

concept that the number of protons determines the identity of an element; the term "atomic number" is then introduced *following* this construction. The "application" of this concept would be to use the periodic table to identify the number of protons characterizing other elements. Alternatively, the activity could begin with groups of students using beads of different colors and sizes to represent protons, neutrons, and electrons in atoms. The students construct atom models by placing the beads in plastic sealed baggies using specific instructions on how many proton, neutron, or electron beads to place in the bags (see Figure 2). In essence, the students produce the "data" that will be explored. Then (possibly the next day), the bags are used to work through an activity similar to Figure 1.

There are two key aspects to the design of any POGIL classroom activity. First, appropriate information must be included for the initial "Exploration," so that students are able to develop the desired concepts. Second, the guiding questions must be carefully constructed and sequenced to enable students to reach the appropriate conclusion, while at the same time encouraging the development of various process skills. Having them reconstruct a table with the data in a certain order, or having them draw a graph and describe the relationship often helps them see patterns in the data more readily. An example involving the use of a pressure probe to investigate the pressure-volume relationship in gases is provided in Figure 3.

The POGIL philosophy is that the development of key process skills (information processing, problem solving, critical thinking, communication, teamwork, self-assessment) is a specific focus of the classroom implementation. POGIL uses course content to facilitate the development of important process skills, including higher-level thinking and the ability to learn and to apply knowledge in new contexts. This approach provides an excellent opportunity to develop most of the key inquiry skills (establishing criteria for evidence; proposing, evaluating, and communicating explanations) in the context of developing content knowledge and investigating questions of scientific interest. Numerous resources are available, providing further information about implementing POGIL (Hanson, 2006; POGIL Project, 2008; Moog et al., 2008).

Although some college courses are taught virtually exclusively using the POGIL approach (Farrell et al., 1999), many high school teachers have found that POGIL activities work best when combined with other methods of instruction. Using POGIL activities as one of a number of classroom techniques can help address the multiple learning styles present in any classroom. For example, one of the authors (Trout) uses a POGIL activity at least once every two weeks, usually at the beginning of a lesson or unit to introduce key concepts. This often provides a strong conceptual foundation, leading to a reduction in the need for review and repetition later in the unit, or the course.

It bears noting that because of the great diversity in student ability, the group interactions of POGIL can present significant challenges. The literature on effective implementation of cooperative and group learning is vast and will not be addressed here. Johnson et al. (1991), Cooper (2005), and Felder (2008) provide excellent resources on this subject, as does the POGIL Instructor's Guide (Hanson, 2006).

Some teachers remain skeptical about using POGIL with high school students. Common remarks include "They don't have enough background knowledge." or "They are not mature enough." In addition, some instructors are concerned about the large quantity of material that must be presented: "I need to cover so much material for standardized testing, I can't afford time for inquiry learning." However, many high school teachers have found that the constructivist approach that POGIL uses is a perfect fit for high school. In fact, the learning cycle structure used in POGIL activities was originally developed for concrete learners in elementary schools. This approach provides students with a solid foundation of scientific thought processes and content. Research has documented the effectiveness of the Learning Cycle Approach in high school science classes (Abraham, 2005) and also the effectiveness of POGIL in a variety of settings (Farrell et al., 1999; Hanson and Wolfskill, 2000; Lewis and Lewis, 2005; POGIL Project, 2008). Teachers who have implemented POGIL in their high school classrooms report great success with difficult topics at basic, regular, and honors levels. Students tend to understand these concepts better, and retain the understanding longer than with previous methods. They are developing the skills of analysis, thought, and communication that are at the heart of inquiry learning. In addition, the students are learning to work as a group, organize information, find patterns, and construct their own, deeper, understanding of concepts.

The Model-Observe-Reflect-Explain Thinking Frame

A second example of a research-based instructional tool that promotes inquiry learning in the chemistry classroom is the Model-Observe-Reflect-Explain, or MORE, Thinking Frame. MORE provides students with a framework for thinking like a chemist engaged in inquiry. Originally designed to be used with multiweek laboratory investigations to facilitate students' successive refinements of their explanations about chemical phenomena (Tien et al., 1999), the MORE Thinking Frame can also be used to transform standard chemistry laboratory experiments and demonstrations into cognitively effective inquiry experiences that incorporate the five essential features of inquiry identified in *Inquiry and the National Science Education Standards* (NRC, 2000).

Using **MORE**, students are first asked to describe their initial understandings (their initial **models**) about the chemical

Figure 3. An activity to investigate Boyle's law.

Investigation of Gas Properties

Samples of gases can be described by several variables, which are all interrelated. We can take the **temperature** of a gas sample, measure the **pressure**, and find the **volume** or the **mass**. As you may have experienced, when you heat a gas sample, its volume changes, or perhaps its pressure. Is there a specific mathematical relationship between these variables? This activity will look at the relationship between volume and pressure specifically, while keeping mass and temperature the same.

Set up a computerized gas pressure probe as instructed. Connect a syringe, about half full of air, to the pressure probe.

1. What is the pressure inside the syringe? What units are you using?

2. Move the plunger on the syringe in and out without creating a leak. Observe the changes in pressure as you do this. Explain **on the molecular level** why the pressure changes.

3. Identify the variables in this activity.

   Independent   Dependent   Controlled

4. Fill in a data table with 10 sets of pressure-volume readings.

   (Data table is provided for students to fill in.)

5. Describe in general the trend or relationship between the variables.

   "As the volume gets smaller, the…."

6. Plot the points of data that you just collected on a sheet of graph paper. Which axis should you label with your independent variable? Which axis is the dependent variable?

7. Scientists often create a model for data using a mathematical relationship. Consider the following types of relationships. What is the basic equation for each? What would a plot of the relationship look like?

   (Hint: Use a graphing calculator to graph each one if you don't remember the shape of the graph.)

   Linear   Inverse   Exponential
   $y = mx + b$

8. Which of the above mathematical models would best fit the plot made with your pressure and volume data?

9. Write an equation, using the variables V and P (instead of x and y) for your data.

Figure 4. Initial model assignment for "The Chemistry of Antacids: How do YOU Spell Relief?" laboratory module, and an example of a high school student's initial model.

Initial Model Assignment

Describe, in words and/or pictures, your understanding of how an antacid works. What do you expect to observe with your senses before and after you (or another person) take an antacid to relieve heartburn or indigestion? Also show how you think an antacid would affect the pH of the stomach contents over time. This is your initial macroscopic model. Then explain what you think the molecules, atoms, and/or ions are doing that results in your observations; this is your initial molecular-level model.

An Example of a High School Student's Initial Model

"An antacid is a more basic substance that will try to neutralize or raise the pH of the acid in the stomachs. The antacids break the acid particles apart to make the molarity lower. When the antacid, if it is a base, is added to the acid in the stomach, they would make water and salt as a product. [Student drawing showing antacid being added to stomach, and water and salt as products.]

Stomach acid (acid) + antacid (base) $\rightarrow$ H_2O (water) + Salt

The smaller the molarity of the acid, the less harsh it will be, Therefore, when the molarity is lowered, it will lower the pH and relieve the pain it is causing.

Important characteristics of effective antacids

pH level, the higher the better; chewable, swallowable, or liquid; molarity, higher pH; size, lower surface area = higher rate of reaction; type of based used. The acid will start out with a high level of hydronium or H_3O^+, and to neutralize it, hydroxide or OH^- must be added. When using an antacid, when it reacts, would it fizz and bubble? Would the fizzing and bubbling have anything to do with LeChatlier's theory?"

system that they will investigate. In these initial models, typically submitted as written prelaboratory assignments, students are encouraged to use words and pictures to describe their understandings from both macroscopic (what students expect to observe and/or measure) and molecular-level (what students think atoms, molecules, and/or ions are doing that would result in the expected observations) perspectives. Pictures are especially useful for communicating molecular-level understandings. (An example of an initial model assignment given to students, and a corresponding student model, is shown in Figure 4.) Student models are then presented and discussed, either in small groups or as a whole class; this makes students aware of alternative understandings and explanations. Next, students gather evidence, typically in the form of experimental observations and/or measurements, which is expected to inform their initial models (**observe**). Third, students monitor the progress of their experiments, seek to understand what is happening, and consider the implications of the evidence being collected as it relates to their initial models (**reflect**). Fourth, students use their evidence to construct a scientific explanation of why their previous model has changed (or why it has not) for presentation to their teacher and other members of the class (**explain**). Following each experiment, students are explicitly prompted to reflect upon the implications of the evidence they have gathered for their model and revise their ideas accordingly (model refinement). Throughout this inquiry process, the essence of chemistry—making connections between macroscopic observations and atomic- and molecular-level explanations—is emphasized. Thus, the MORE Thinking Frame provides cognitive guidance and support for students as they propose, communicate, evaluate, and refine their own evidence-based explanations to address scientifically oriented questions. It provides students with a framework for *thinking like a chemist* engaged in inquiry, in contrast to traditional laboratory exercises that typically focus on providing students with instructions for carrying out physical manipulations in the laboratory.

By virtue of constructing and refining their models in light of the evidence that they gather, students using the MORE Thinking Frame engage in all five of the essential features of inquiry learning, including investigating scientifically oriented questions, proposing explanations, establishing criteria for evidence, and evaluating and communicating explanations. In addition, the MORE Thinking Frame combines a focus on metacognition (thinking about one's own thinking) with many interrelated elements that research has found to be among the most effective for enhancing science learning. These elements include activating students' prior knowledge (Alvermann and Hynd, 1989; Marazano et al., 2001), encouraging students to combine linguistic and nonlinguistic modes to represent their understanding (Marazano et al., 2001; Mayer, 1989), promoting testing and revision of models (Marazano et al., 2001; White, 1993, 1998), fostering cognitive conflict or dissatisfaction with naïve conceptions (Guzzetti et al., 1993), and scaffolding students' engagement in authentic scientific thinking processes (Brown and Campione, 1994; Collins et al., 1989).

Implementing the MORE Thinking Frame in High School Chemistry Classrooms

Implementing MORE in the high school chemistry classroom can be very rewarding, for both the student and the teacher, and at the same time, somewhat challenging for first-time practitioners. While being extremely beneficial to learning, the MORE framework takes additional time to both implement in the classroom and to check for student understanding (grading). In the beginning, instructors may find implementation challenging, but after sufficient experience with it, they often question whether their students really ever learned without it. Using the MORE method in the laboratory is so successful that instructors often start to use aspects of it in their "lectures", and the traditional separation between lecture and laboratory blurs. Instructors begin to ask students higher-level questions that require reflection upon and explanations of what students think is happening on the molecular level. Teachers begin to spend less time presenting what they know of chemistry to students, and become more concerned with what and how their students think. When instructors implement MORE, learning starts to happen on a deeper level for both students and teachers.

To compare and contrast traditional laboratory methods with MORE, we explore the differences between a typical acid-base laboratory experiment and a MORE laboratory module focused on constructing evidence-based explanations of how antacids work. In the traditional acid-base laboratory experiment, students typically add acid-base indicators to various substances, including various solutions from the stockroom, as well as household chemicals. Students then record data into their lab notebooks and answer a set of questions relating to the substances' pHs (if any questions are present at all). Students are graded on how close they come to the solutions' accepted pHs and how well they answered the required questions. An extension might be added later, in which the students use titrations to figure out the concentration of an unknown acid or base solution.

A MORE laboratory investigation that is intended to foster an understanding of the concepts and principles of acid-base chemistry looks quite different than the traditional one, and the resulting learning outcomes are very different too. For example, in a MORE laboratory module entitled "The Chemistry of Antacids: How do YOU Spell Relief?", students are first asked to propose an explanation for how antacids work. In these initial models, students are asked to explain what they think they will observe on the *macroscopic* level when they add antacids to the stomach, and to provide their ideas about what happens on the *molecular* level that explains these observations. (See Figure 4 for the full initial model assignment.) A whole-class discussion of the students' initial models then follows, allowing students the opportunity to communicate their explanations and compare their ideas with those of other students.

After the students discuss their initial models, they are guided through investigations such as finding the change in pH when adding an antacid to a simulated stomach, relative solubilities of different antacids and effects of solubility on pH, and determining the antacid-neutralizing capacities of the different antacids. When conducting their experiments (which are primarily designed by the students themselves), students are encouraged to discuss their findings with their peers and to reflect on how the evidence they have gathered relates to their initial ideas. To stimulate this student reflection, the teacher poses questions to groups of students. Some general questions that we have found effective include

- What is the goal of this experiment?
- How does what you are doing contribute to the goal of the experiment?
- How does what you are currently observing relate to your initial model?
- Are your observations consistent with your initial model? Explain.
- Does your model fully explain your observations? How?
- What do you think is happening on the molecular level?
- What doesn't make sense to you?

After each experiment, students evaluate their initial explanations in light of the experimental evidence they have collected, refine their models of antacids based on what they have observed, and discuss their refinements with the class. After several cycles of experimentation, students write a final refined model that not only includes their final explanation of how antacids work, but also information about each refinement they have made

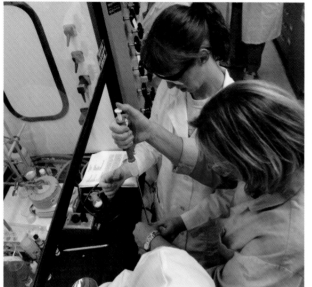

David Armer, USNCO

to their models along the way. (See Figure 5 for excerpts from the final refined model for the student whose initial model appears in Figure 4.) The students' models are not assessed on how scientifically correct their explanations have become, but rather, on how consistent their claims are with the evidence they have gathered. An important point to note is that to encourage this kind of work from students effectively, instructors must move away from thinking that students ideas always need to be *fully* correct. Rather, MORE instruction emphasizes the process of students constructing explanations that are consistent with the evidence they have gathered, which should ultimately lead to understandings consistent with the scientifically accepted views. As you can see from Figures 4 and 5, the student productively refines their model of how antacids work, both from macroscopic and molecular-level perspectives, but their molecular-level views are not yet fully correct by the time they write their final model. Clearly, teachers obtain important feedback about students' developing understandings from reading their students' models that they would not be able to obtain from reading traditional laboratory reports.

In addition to presenting the refined model (from both macroscopic and molecular-level perspectives), research on the use of the MORE Thinking Frame has shown that it is very important for students' understanding of the chemistry concepts for them to explicitly *explain why their model has changed* (or why it has not if the experimental evidence supports the initial model), using specific experimental evidence. For example, in the refined model shown in Figure 5, the student wrote, "We discovered that the most effective antacids have calcium carbonate as their main active ingredient. This means it won't create a salt and water but instead it will create water, carbon dioxide, and a salt is created. This would explain why the product bubbles and fizzes." Although we would like to see the model expressed in a bit more detail, this student is explaining that their ideas changed from thinking that an antacid must contain hydroxide (OH⁻) and wondering whether that would lead to the macroscopic observation of fizzing and bubbling (see Figure 4) to understanding that an active ingredient in many antacids is carbonate (CO_3^{2-}), which is consistent with the observation of bubbles of carbon dioxide gas when the antacid is added to acid. These model changes were based on gathering data from antacid labels and from observations made when adding various antacids to aqueous acid solutions. Thus, the student is using evidence to refine their explanation of how antacids work and communicate an awareness of how their ideas changed as a result of participating in scientific inquiry.

Research on the use of the MORE Thinking Frame in high school chemistry classes reveals that students participating in MORE laboratory experiences outperform control groups participating in more traditional laboratory experiences on written chemistry examinations. In addition, preliminary video analyses indicate that two main aspects of MORE activities prompted student molecular-level discussions at the high school level: classroom model writing and instructor questioning during the experimental portion of the class period (Carillo et al., 2005). Student survey and interview data also show that, in contrast to students in standard classes, students participating in MORE inquiries learned to value thinking about what is happening on the molecular level to explain their observations.

Implementing Inquiry *Effectively* in the High School Classroom

The experience of teaching chemistry through inquiry, and what is required of the teacher, is very different compared with teaching via traditional methods, in part because the students' ideas often dictate the direction that class discussions and experimental investigations will take. Teachers using inquiry in the classroom need to constantly adjust pacing and follow the students' lead to some extent, not simply proceed through lecture notes. For many instructors, the change to inquiry instruction from more traditional approaches provides numerous challenges, particularly for those who have not experienced inquiry as a student or had training in its effective implementation.

Even though there is a great deal of research indicating that inquiry approaches can be very effective, this does not guarantee that every attempt at inquiry will be successful. In fact, some instructors may be leery of implementing inquiry because of prior experiences in which the "recommended" approach was to provide students with access to various materials and "let them loose" to explore on their own - in hopes that they would "discover" some pattern or scientific principle and be able to explain it. This approach frequently leads to arguments against inquiry from teachers: "I can't afford the time." "Students don't have the natural ability they need for inquiry." "Students don't have the necessary background." The problem is that the inquiry experiences on which these opinions are based likely were not constructed with enough student guidance to achieve the goals that were intended.

For many instructors, the logical place to consider implementing inquiry instruction is in the context of a laboratory setting. Laboratory-based projects (such as a science fair investigation), in which the student independently selects the topic and the question to examine, provide the typical example of an open-inquiry experience. Many teachers' reaction to inquiry-based labs is disdain for the chaos they create in the classroom. Using a guided approach in which the students are asked to formulate a question within a topic, design an experimental protocol to gather evidence to address their question, and construct an evidence-based explanation of their results offers a happy medium. For example, if the topic is kinetics, the students can be asked to hypothesize what variable they might alter to increase the rate of a particular reaction. Students will come up with several different questions and procedures, but the instructor only needs to prepare one system of reactants and materials. There may be some extra preparation required to set up equipment (for example, one group might need a hot plate, while another needs ice), but the chaos is limited. While many instructors have found that laboratory experiments take more time, guided inquiry ultimately provides better educational outcomes when students plan (at least part of) the procedure themselves, and perhaps ask their own questions, but the focus is on guiding students in proposing, evaluating (based on the evidence they collect), and communicating a scientific explanation.

Inquiry learning need not be limited to the realm of the laboratory. For example, data or manipulatives may be provided to the students in place of results from a laboratory investigation. If appropriate information is provided and is accompanied by a carefully crafted

Figure 5. Excerpts from student's final refined model for "The Chemistry of Antacids: How do YOU Spell Relief?" laboratory module. (Same student whose initial model is shown in Figure 4.)

An Example of a High School Student's Final Refined Model

"On the macroscopic level the main thing we can see is fizzing and bubbling. The color also changed because of the pH indicator. When we started, the acid had a very low pH and the antacid always significantly raised the pH … On the molecular level, when the basic molecules come into contact with the acidic ones, they break each other apart and form a substance. We discovered that the most effective antacids have calcium carbonate as their main active ingredient. This means it won't create a salt and water but instead it will create water, carbon dioxide, and a salt. This would explain why the product bubbles and fizzes. [Student drawing showing bubbles labeled "CO_2," rising from a container of liquid. The liquid phase is labeled "H_2O + salt."] I am not sure what the actual antacid particles look like on a molecular level, but I am guessing the antacid particles are attracted to the acid particles to even the pH out. The antacid particles come between the acid particles and reattach with their polar matched ion. [Student drawing showing reactants composed of Ant^+Ant^- and H^+Cl^- and products composed of Ant^+Cl^- and H^+Ant] These antacids don't necessarily contain hydroxide but more often contain a type of carbonate. It is true that the lower the original molarity of the acid, the easier and faster the pH will raise. We were unable to test the pH of the antacid before, so we don't know the original pH of the antacids …"

set of questions leading the students through the logical progression necessary to understand a concept, then the inquiry learning experience can be powerful, particularly in developing the students' abilities in proposing, evaluating, and communicating logical and evidence-based explanations of their thinking. The activity described previously in Figure 2 provides an example of this type of approach.

A number of approaches to inquiry instruction for the physical sciences at the high school and introductory college levels that have been classroom-tested and shown to be effective are now widely used. Among these are the Physics by Inquiry program from the University of Washington (McDermott, 1996); the Modeling Instruction Program for physics, chemistry, and physical sciences from Arizona State University (Modeling Instruction Program, 2008); Living by Chemistry from the University of California at Berkeley (Stacy, 2008); the Science Writing Heuristic approach to laboratory experiences and report writing (Greenbowe and Hand, 2005); and the Discovery Chemistry curriculum from the College of the Holy Cross (Ditzler and Ricci, 1991, 1994; Ricci et al., 1994).

David Armer, USNCO

Incorporating inquiry-based instructional methods, such as Process-Oriented Guided Inquiry (POGIL) and the Model-Observe-Reflect-Explain (MORE) Thinking Frame, in the high-school chemistry classroom facilitates students' learning of how to investigate chemical problems, as well as understanding the process and nature of chemistry, leading to more robust understandings of chemistry content.

Recommended Readings

Carillo, L.; Lee, C.; Rickey, D. Enhancing Science Teaching by doing MORE: A Framework to Guide Chemistry Students' Thinking in the Laboratory. *Sci. Teach.* (Special Issue: Inquiry in the Laboratory) 2005, *72*, 60–64. The article discussing the use of the MORE Thinking Frame in high school chemistry classes provides additional examples to supplement those presented here.

Moog, R. S.; Creegan, F. J.; Hanson, D. M.; Spencer, J. N.; Straumanis, A. R. Process-Oriented Guided Inquiry Learning. *Metro. Univ. J.* 2006, *17*, 41–51. This article provides a general overview of POGIL and contains many additional references.

National Research Council (NRC). *Inquiry and the National Science Education Standards;* National Academies Press: Washington, DC, 2000. This addendum to the NSES focuses on the essential features of integrating inquiry learning into science instruction.

Recommended Web Sites

http://more.colostate.edu/ This Web site provides information about using the Model-Observe-Reflect-Explain (MORE) Thinking Frame, including contact information for developers and implementers of MORE. (accessed March 2008).

http://www.pogil.org/ This Web site provides information about Process-Oriented Guided Inquiry (POGIL), including data about the effectiveness of the approach, a downloadable version of the Instructor's Guide to POGIL, and information on upcoming workshops and other events. (accessed March 2008).

References

Abraham, M. R. The learning cycle approach as a strategy for instruction in science. In *International Handbook of Science Education;* Tobin, K., Fraser, B., Eds.; Kluwer: The Netherlands, 1998; pp 513–524.

Abraham, M. R. Inquiry and the learning cycle approach. In *Chemists' Guide to Effective Teaching;* Pienta, N. J., Cooper, M. M., Greenbowe, T. J., Eds.; Prentice Hall: Upper Saddle River, NJ, 2005.

Abraham, M. R.; Renner, J. W. Research on the learning cycle. *J. Res. Sci. Teach.* 1986, 23, 121–143.

Alvermann, D. E.; Hynd, C. R. Effects of prior knowledge activation modes and text structure on nonscience majors' comprehension of physics. *J. Educ. Res.* 1989, *83,* 97–102.

Bodner, G. M. I have found you an argument: The conceptual knowledge of beginning chemistry graduate students. *J. Chem. Educ.* 1991, *68,* 385–388.

Brown, A. L.; Campione, J. C. Guided Discovery in a Community of Learners. In *Classroom Lessons: Integrating Cognitive Theory and Classroom Practice;* McGilly, K., Ed.; MIT Press: Cambridge, MA; 1994; pp 229–270.

Carillo, L.; Lee, C.; Rickey, D. See Recommended Readings.

Collins, A.; Brown, J. S.; Newman, S. E. Cognitive Apprenticeship: Teaching the Crafts of Reading, Writing, and Mathematics. In *Knowing, Learning, and Instruction: Essays in Honor of Robert Glaser;* Resnick, L. B., Ed.; Lawrence Erlbaum Associates: Hillsdale, NJ, 1989; pp 453–494.

Cooper, M. M. An introduction to small group learning. In, *Chemists' Guide to Effective Teaching*; Pienta, N. J., Cooper, M. M., Greenbowe, T. J., Eds.; Prentice Hall: Upper Saddle River, NJ:, 2005.

Cros, D.; Amouroux, R.; Chastrette, M.; Fayol, M.; Leber, J.; Maurin, M. Conceptions of first-year university students of the constituents of matter and the notions of acids and bases. *Eur. J. Sci. Educ.* 1986, *8,* 305–313.

Cros, D.; Chastrette, M.; Fayol M. Conceptions of second-year university students of some fundamental notions in chemistry. *Int. J. Sci. Educ.* 1988, *10,* 331–336.

Ditzler, M. A.; Ricci, R. W. Discovery chemistry: Balancing creativity and structure. *J. Chem. Educ.* 1994, *71,* 685–688.

Ditzler, M. A.; Ricci, R. W. Discovery chemistry: A laboratory-centered approach to teaching general chemistry. *J. Chem. Educ.* 1991, *68,* 228–232.

Farrell, J. J.; Moog, R. S.; Spencer, J. N. A guided inquiry general chemistry course. *J. Chem. Educ.* 1999, *76,* 570–574.

Felder, R. M. Publications on Cooperative Learning. http://www4.ncsu.edu/unity/lockers/users/f/felder/public/Student-Centered.html#Publications-Coop, 2008. (accessed March 2008).

Gabel, D. L.; Samuel, K. V.; Hunn, D. J. Understanding the particulate nature of matter. *J. Chem. Educ.* 1987, *64,* 695–697.

Greenbowe, T. J.; Hand B. M. Introduction to the science writing heuristic. In *Chemists' Guide to Effective Teaching.* Pienta, N. J., Cooper, M. M., Greenbowe, T. J., Eds.; Prentice Hall: Upper Saddle River, NJ, 2005.

Gunstone, R. F.; White, R. T. Understanding of gravity. *Sci. Educ.* 1981, *65,* 291–299.

Guzzetti, B. J.; Snyder, T. E.; Glass, G. V.; Gamas, W. S. Promoting conceptual change in science: A comparative meta-analysis of instructional interventions from reading education and science education. *Read. Res. Quart.* 1993, *28,* 117–159.

Hanson, D. M. *Instructor's Guide to Process Oriented Guided Inquiry Learning.* Pacific Crest: Lisle, IL, 2006. Also available as a download from the POGIL Web site at http://www.pogil.org/resources/pogil_ig.php (accessed March 2008).

Hanson, D. M; Wolfskill, T. Process workshops: A new model for instruction. *J. Chem. Educ.* 2000, *77*, 120–130.

Hofstein, A. The laboratory in chemistry education. *Chem. Educ. Res. Pract.* 2004, *5*, 247–264.

Hofstein, A.; Lunetta, V. N. The role of the laboratory in science teaching: Neglected aspects of research. *Rev. Educ. Res.* 1982, *52*, 201–217.

Johnson, D. W.; Johnson, R. T.; Smith, K. *Active Learning: Cooperation in the College Classroom*, Interaction Book Company: Edina, MN, 1991.

Lawson, A. E.; Abraham, M. R.; Renner, J. W. *A Theory of Instruction:Using the Learning Cycle to Teach Science Concepts and Thinking Skills* [Monograph, Number one]. National Association for Research in Science Teaching: Kansas State University, Manhattan, KS, 1989.

Lawson, A. E. *Science Teaching and the Development of Thinking*. Wadsworth Publishing Company: Belmont, CA, 1995.

Lawson, A. E. What should students know about the nature of science and how should we teach it? *J. Coll. Sci. Teach.* 1999, 28, 401–411.

Lazarowitz, R.; Tamir, P. Research on using laboratory instruction in science. In *Handbook of Research on Science Teaching and Learning*; Gabel, D. L., Ed.; Maxwell Macmillan International: New York, 1994; pp 94–128.

Lewis, S. E.; Lewis, J. E. Departing from lectures: An evaluation of a peer-led guided inquiry alternative. *J. Chem. Educ.* 2005, *82*, 135–139.

Marazano, R. J.; Pickering, D.; Pollock, J. E. *Classroom Instruction that Works: Research-Based Strategies for Increasing Student Achievement*; Association for Supervision and Curriculum Development: Alexandria, VA, 2001.

Mayer, R. E. Models for understanding. *Rev. Educ. Res.* 1989, *59*, 43–64.

McDermott, L. C. *Physics by Inquiry.* New York: John Wiley & Sons, 1996; Vol. I and II.

Modeling Instruction Program. Available online at http://modeling.asu.edu (accessed March 2008).

Moog, R. S.; Creegan, F. J.; Hanson, D. M.; Spencer, J. N.; Straumanis, A. R.; Bunce, D. M.; Wolfskill, T. POGIL: Process-Oriented Guided Inquiry Learning. In *Chemists' Guide to Effective Teaching;* Pienta, N. J.; Cooper, M. M.; Greenbowe. T. J., Eds. Pearson Prentice Hall: Upper Saddle River, NJ, 2008; Vol. 2.

Moog, R. S.; Creegan, F. J.; Hanson, D. M.; Spencer, J. N.; Straumanis, A. R. Process-oriented guided inquiry learning. *Metro. Univ. J.* 2006, *17*, 41–51.

Nakhleh, M. B. Why some students don't learn chemistry. *J. Chem. Educ.* 1992, *69*, 190–196.

National Research Council (NRC). *Inquiry and the National Science Education Standards;* National Academies Press: Washington, DC, 2000.

NRC. *National Science Education Standards*; National Academies Press: Washington, DC, 1996.

Panitz, T. The Case for Student Centered Instruction Via Collaborative Learning Paradigms. Available online at http://home.capecod.net/~tpanitz/tedsarticles/coopbenefits.htm (accessed March 2008).

Papert, S. A. *Mindstorms: Children, Computers, and Powerful Ideas*; Basic Books: New York, 1980.

POGIL Project. Effectiveness of POGIL. Available online at http://www.pogil.org/effectiveness/ (accessed March 2008).

Riah, H.; Fraser, B. J. Chemistry Learning Environment and Its Association with Students' Achievement in Chemistry. Presented at the Annual Meeting of the American Educational Research Association, San Diego, CA, 1998.

Rickey, D.; Stacy, A. M. The role of metacognition in learning chemistry. *J. Chem. Educ.* 2000, *77*, 915–920.

Rickey, D.; Teichert, M. A.; Tien, L. T. Model-Observe-Reflect-Explain (MORE) Thinking Frame Instruction: Promoting Reflective Laboratory Experiences to Improve Understanding of Chemistry. In *Chemists' Guide to Effective Teaching*; Pienta, N. J., Cooper, M. M, Greenbowe, T. J., Eds.; Pearson Prentice Hall: Upper Saddle River, NJ, 2008; Vol 2.

Ricci, R. W.; Ditzler, M. A.; Jarret, R.; McMaster, P.; Herrick, R. The Holy Cross discovery chemistry program. *J. Chem. Educ.* 1994, *71*, 404–405.

Rund, J. V.; Keller, P. C.; Brown, S. L. Who does what in freshman lab? A Survey. *J. Chem. Educ.* 1989, *66*, 161–164.

Smith, K. J.; Metz, P. A. Evaluating student understanding of solution chemistry through microscopic representations. *J. Chem. Educ.* 1996, *73*, 233–235.

Stacy, A. *Living by Chemistry*. Key Curriculum Press: Emeryville, CA, 2008. Available online at http://www.keypress.com/x4716.xml (accessed March 2008).

Tien, L. T.; Rickey, D.; Stacy, A. M. The MORE thinking frame: Guiding students' thinking in the laboratory. *J. Coll. Sci. Teach.* 1999, *28*, 318–324.

Tien, L. T.; Teichert, M. A.; Rickey, D. Effectiveness of a MORE laboratory module in prompting students to revise their molecular-level ideas about solutions. *J. Chem. Educ.* 2007, *84*, 175–181.

Vye, N. J.; Schwartz, D. L.; Bransford, J. D.; Barron, B. J.; Zech, L. SMART environments that support monitoring, reflection, and revision. In *Metacognition in Educational Theory and Practice*; Hacker, D. J., Dunlosky, J., Graesser, A. C., Eds.; Lawrence Erlbaum Associates: Mahwah, NJ; 1988; pp 305–346.

White, B. Y. Thinkertools: Causal Models, Conceptual Change, and Science Education. *Cognit. Instr.* 1993, *10*, 1–100.

White, B. Y.; Frederickson, J. R. Inquiry, Modeling, and Metacognition: Making Science Accessible to All Students. *Cognit. Instr.* 1988, *16*, 90–91.

White, R. T. Implications of recent research on learning for curriculum and assessment. *J. Curr. Stud.* 1992, *24*, 153–164.

Chapter 5

Chemistry and the Life Science Standards

by Deborah Herrington, Ellen Yezierski, and Rebecca Caldwell

Deborah Herrington *graduated from the University of Waterloo with an M.Sc. in chemistry and from Purdue University with a Ph.D. in chemistry education. She has been an assistant professor of chemistry at Grand Valley State University in Allendale, MI, for the past 4 years. Debbie is the codirector of Target Inquiry, an innovative professional development program for high school chemistry teachers. Contact e-mail: herringd@gvsu.edu*

Ellen Yezierski, *a former high school teacher, graduated from Arizona State University with a Ph.D. in Curriculum and Instruction (Science Education). She is currently an Assistant Professor of Chemistry at Grand Valley State University in Allendale, MI. Ellen codirects Target Inquiry (TI), a new professional development program for high school chemistry teachers, and is currently studying TI's impacts on teachers and their students. Contact e-mail: yezierse@ gvsu.edu*

Rebecca Caldwell *graduated from the University of Michigan–Dearborn with a degree in secondary chemistry and math education. She continued her education and earned a graduate degree from Purdue University in chemical education. She has taught high school chemistry for the past six years at Trenton High School in Trenton, MI. Contact e-mail: caldwelr@trenton. k12.mi.us*

Students have a genuine interest in biological processes, as they are relevant to their daily lives. Students also have first-hand experience with many biological concepts. Therefore, using biological examples in chemistry classrooms can not only tap into students' inherent interest in biology, but also help them make important interdisciplinary connections. Additionally, *No Child Left Behind* and state science standards expect *all* students to have a more comprehensive and connected understanding of science.

However, the *National Science Education Standards* (NSES) suggest that many students associate molecules with physical science and fail to understand that living systems are composed of molecules. The inability of students to make connections between important biological processes and key chemistry concepts such as atoms and molecules, structure and properties of matter, thermodynamics, and chemical reactions is largely the result of how we teach chemistry. We want our students to understand that chemistry is relevant to their everyday lives, yet many instructional examples involve chemicals or contexts with which students have no practical experience. This makes it difficult for students to construct the desired connections to other disciplines and encourages them to confine chemistry concepts to the chemistry classroom. Moreover, research indicates the transfer of information from one context to another is enhanced by presentation of material in multiple contexts (Bransford, 2000).

Therefore, it is notable that in the NSES, there is not listed any chemistry standards. Although most teachers clearly recognize that many of the physical science standards relate specifically to chemistry, there are also a number of life science standards that incorporate chemistry concepts. The overlap between the chemistry concepts found in the physical and life science standards is shown in Table 1.

Table 1: Chemistry in the Life Science and Physical Science Standards

Physical Science Standards	Related Life Science Standards
Structure of atoms Structure and properties of matter Chemical reactions Conservation of energy	**The Cell** Cells are made up of molecules Different types of molecules form specialized cell structures (e.g., membranes) Most cell functions involve chemical reactions Cells use chemical reactions to produce energy
Structure and properties of matter	**Molecular basis of heredity** Chemical and structural properties of DNA explain how genetic information is encoded and replicated
Conservation of energy and increase in disorder Chemical reactions Interactions of matter and energy	**Matter, energy, and organization in living systems** All matter tends toward more disorganized states Plants capture energy from the sun to form covalent chemical bonds ATP is a high-energy molecule

Table 2 summarizes changes that will promote the integration of chemistry and the life sciences. The remainder of the chapter discusses each content area in Table 2 in detail, including annotated curricular materials and highlighted classroom examples. This does not require teaching a separate course in organic chemistry or biochemistry, but rather linking certain key chemistry concepts to life science teaching.

Table 2: Changing Emphases To Integrate Chemistry and the Life Sciences

Content Area	Less emphasis on	More emphasis on
Structure and shape	primarily inorganic examples VSEPR to teach shape with no discussion of context	Lewis structures of organic molecules and organic functional groups biological importance of molecular shape
Intermolecular forces and physical properties	small-molecule examples of intermolecular forces	biological examples of intermolecular forces
Chemical reactions	reactions and problems involving reagents unfamiliar to students	reaction and stoichiometry problems with biological significance and small organic molecules (such as acetic acid)
Thermodynamics	calorimetry examples with metals and water Hess's law examples with inorganic reactions	food metabolism examples and labs biological examples of exothermic and endothermic reactions Hess's law examples with small organic molecules

Structure and Shape

Generally, lessons on ionic and covalent bonding are followed by a discussion of the structure and shapes of molecules. Students typically learn about the octet rule and how to draw Lewis structures, as well as how to use Lewis structures to determine the three-dimensional shapes of molecules using Valence Shell Electron Pair Repulsion (VSEPR) theory. Traditionally, we use small inorganic molecules to teach these concepts because students can focus on just a few atoms. However, the majority of covalent compounds that students will encounter in their lives are organic molecules; the structure and shape of organic molecules play an important role in many biological processes.

One example in which the structure and shape play an important role is in smell. The structure of organic molecules impacts smell as different organic functional groups tend to have different types of odors. For example, amines generally have a fishy smell, esters have a sweet fruity smell, and carboxylic acids have a pungent or putrid smell. Figure 1 shows two organic molecules with the same molecular formula but different functional groups. These two compounds have very different odors.

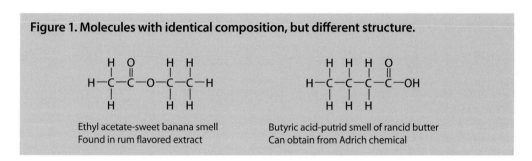

Figure 1. Molecules with identical composition, but different structure.

Ethyl acetate-sweet banana smell
Found in rum flavored extract

Butyric acid-putrid smell of rancid butter
Can obtain from Adrich chemical

The *Living by Chemistry* smells curriculum (Stacy et al., 2003*a*) uses smells to introduce students to organic molecules. By classifying compounds based on smell and comparing the molecular formulae of similar smelling compounds, students are able to relate organic functional groups to particular smells. Furthermore, by examining a number of organic structures, students begin to develop the rules for drawing organic structures: H atoms form one bond, O atoms 2, N atoms 3, and C atoms 4. They call this the "HONC—1234 rule." Connections can then be made to the periodic table, with the number of valance electrons providing the basis for drawing Lewis structures of both organic and inorganic molecules.

However, it is not just the functional group that determines the smell of a compound; the shape of a molecule also affects its odor. Similar to the way that substrate molecules fit into enzyme-active sites like a key fits into a lock, smell receptors also bind molecules of a specific shape. Discovering the relationship between odor and shape can lead to a discussion of VSEPR and three-dimensional shape. Students can begin by making models of small molecules such as CH_4, NH_3, H_2O, CO_2, C_2H_6, CH_3OH, etc., to clearly see how the number of bonds and lone pairs around the central atom affect the three-dimensional shapes of molecules. Finally, students can either build or examine prebuilt models of larger organic molecules to make connections between shape and smell. For example, sweet-smelling compounds tend to be linear or "stringy," flat molecules have a minty odor, and more spherical molecules smell like camphor (Stacy et al., 2003*b*). Organic molecules provide excellent examples of molecular shapes up to tetrahedral. Going beyond tetrahedral electron pair geometries is usually unnecessary, as students will not encounter molecules with 5 or 6 regions of electrons around the central atom until more advanced college chemistry courses.

Purpose: To demonstrate the effect of molecular shape on taste

Background: Most students are familiar with the concept of photosynthesis—plants make glucose from CO_2, H_2O, and light—but few have any real understanding of glucose on the molecular level. Additionally, students may not know that there are different kinds of sugars and that sugar units (monosaccharides) link together to form larger sugars or carbohydrates (Institute of Food Technologies, 2007). Glucose and fructose are both sugars with the chemical formula $C_6H_{12}O_6$; however, fructose is much sweeter than glucose. Sucrose is used as the standard for measuring sweetness and is given the value of 1. Fructose has a sweetness value of 170 and glucose 80 (Ontario Ministry of Agriculture, 2007). The sweetness of a sugar is largely dependent on its shape (Emsley, 1988; Purves, 2006).

Activity: After students have learned about VSEPR and three-dimensional shape, have them build models of fructose and glucose to examine how two molecules with the same molecular formula can have different three-dimensional shapes. Students can also connect the models of glucose and fructose to see how sucrose (table sugar) is formed. Structures can be found on the Web (Lyndaker, 2007).

Assessment: Students can be asked the following questions: (1) What is similar and what is different about the structures of glucose and fructose? (2) How could these structural differences affect how sweet it tastes? (3) What is removed when you connect glucose and fructose together to form sucrose?

Extensions: Students can build a model of an artificial sweetener, aspartame, and compare its structure to sucrose and fructose. The structure of aspartame can be found at http://www.3dchem.com.

Intermolecular Forces and Physical Properties

Typically, after molecular shape, students learn about polarity and intermolecular forces, as well as how intermolecular forces impact physical properties, such as melting point and solubility. However, we often forget that intermolecular forces play a very important role in the life sciences. For example, polarity plays an important role in vitamin solubility. Polar vitamins (like vitamin C) are water soluble, whereas nonpolar vitamins (like vitamin E) are fat soluble. An example of the important role that lipids (nonpolar molecules) play in the body is found in the *ChemMatters* student activity "The Aspirin Effect: Pain Relief and More" (Kimbrough, 2004). In reading this article, students learn about the role of prostaglandins in pain and swelling and how drugs like aspirin, ibuprofen, and acetaminophen (Tylenol) work to inhibit the prostaglandin synthesis pathway. An important connection to chemistry is that aspirin and ibuprofen work similarly and contain similar functional groups, but acetaminophen has different functional groups and inhibits a different part of the synthesis pathway. The teacher's guide for this activity provides additional instructional materials, including background information, student assessment questions, and structured student reading guides, as well as a table connecting each activity to the NSES.

Noncovalent interactions are also important in DNA and protein structure. The DNA double helix is held together by hydrogen bonds; adenine always pairs with thymine and cytosine with guanine because these pairs have maximal hydrogen bonding interactions within the confined geometry of the double helix. Hydrogen bonds between amino acid amide groups are also what hold protein alpha helices and beta sheets together. The three-dimensional folding of proteins depends on both hydrogen bonding and nonpolar interactions. Often, the nonpolar amino acid side chains are found on the inside of the protein, forming a hydrophobic core, and the polar groups are found on the outside of the protein, helping to make the protein soluble in water. Polar and nonpolar interactions are also important for enzyme substrate specificity. Both the shape of the active site and noncovalent interactions between the amino acid side chains in the active site and the substrate ensure that only specific molecules will fit into an active site. Molecular structures of DNA, DNA base pairs, alpha helices, beta sheets, and small proteins can be found on the Web (Ophardt, 2003).

More important than just placing a chemistry concept into a biological context, using biological examples may help address a common student misconception. Despite emphasizing the difference between covalent bonds and intermolecular forces and stressing that when water boils or sugar dissolves, bonds are not being broken, many students hold onto the belief that for a liquid to enter the gas phase or when a covalent compound dissolves in water, it must break apart. In fact, some entering chemistry graduate students still have this misconception (Bodner, 1991). However, challenging this misconception with a biological example can help students develop the correct scientific explanation. *Living by Chemistry* again uses the concept of smells to do this (Stacy et al., 2003b). Students can understand that although the substances they smell are liquids, to travel to their nose through the air, they must become gases. To address the question of whether molecules break up or remain whole when they leave the liquid state and become a gas, students are presented with four pictorial models: (1) atoms in the liquid state and atoms in the gas state, (2) molecules in the liquid state and molecules in the gas state, (3) atoms in the liquid state and molecules in the gas state, and (4) molecules in the liquid state and atoms in the gas state. Students are then guided through an activity comparing butyric acid and ethyl acetate (see Fig. 1), testing the models to see which best accounts for the observed physical data. This strategy is similar to that of the Process-Oriented Guided Inquiry Learning activities described in chapter 4 of this book.

Chemical Reactions and Stoichiometry

The study of chemical reactions usually focuses first on the qualitative evidence of a chemical reaction (color change, gas evolution, temperature change, precipitate formation, etc.) followed by writing balanced chemical equations for reactions; categorizing reactions into four or five types (synthesis, decomposition, single replacement, double replacement, and combustion of hydrocarbons); and finally, introducing coefficients in chemical equations as mole ratios (stoichiometry). As inorganic compounds contain relatively few atoms or ions, they are ideally suited for this core qualitative and quantitative approach to chemical reactions that is central to chemistry. However, inorganic reactions often use elements and compounds that are unfamiliar to students. Using biologically relevant examples can provide students with a familiar context as well as draw upon their natural interest in biology.

Focusing on particular inorganic substances that affect (or may be found in) the human body and how quantities of these substances are measured provides a more biologically focused approach to studying chemical reactions. Many inorganic ions and compounds have biological connections, such as calcium phosphate (found in bones), lead ions (toxic to humans), and calcium oxalate (one type of kidney stone). Common blood tests such as the CHEM-20 include tests for serum (blood) calcium and chloride ions. These ions are precipitated with oxalate and silver ions, respectively. These examples can be used in stoichiometry problems and laboratory experiments. An online medical encyclopedia contains normal levels of these ions in healthy human blood to provide a realistic and health-focused context for the problems (Medline Plus, 2005). Addressing stoichiometry problems involving serum can also be used to introduce and practice net ionic equations, as well as solution stoichiometry. The ions and compounds mentioned above can be seamlessly added to precipitation labs and problem sets. Here is an example of how a straightforward stoichiometry exercise can be transformed into a problem with a biological context:

Old exercise: If 1.50 g of magnesium iodide reacts with an excess of sodium carbonate, what mass of magnesium carbonate will be formed, according to the equation below?

$$MgI_2 \text{ (aq)} + Na_2CO_3 \text{ (aq)} \rightarrow MgCO_3 \text{ (s)} + 2NaI \text{ (aq)}$$

Kool-Aid Chromatography

Purpose: To use chromatography to separate food dyes based on polarity

Background: Chromatography separates compounds based on the attraction between the compound and solvent (mobile phase) compared to that between the compound and the solid column (stationary phase). Artificial food dyes are complex organic molecules with a number of different functional groups (Schneider et al., 2007). The differences in the molecular structures affect the polarity of the molecules and, in turn, their attraction for the mobile phase versus the stationary phase. Dyes more attracted to the mobile phase will more quickly move through the column, whereas dyes more attracted to the stationary phase will better adsorb on the column and take longer to elute.

Activity: Use chromatography to separate the dyes in a grape drink such as Kool-Aid. Pretreat the Sep-Pak C18 column with 70% isopropyl alcohol and deionized water. Load the column by injecting it with grape drink from a 10-ml syringe. Note the color of the eluent. Repeat with 5% and 25% isopropyl alcohol noting the color of the eluent each time (University of Minnesota, 2000). Sep-Pak C18 cartridges, including instructions for this activity, may be purchased from Flinn Scientific.

Assessment: Ask students the following questions: (1) What serves as the stationary and mobile phases in this experiment? (2) Which eluant is the most polar? least polar? How do you know? (3) Which dye, red or blue, had the most attraction for the column? (4) Which dye had the strongest attraction for the isopropyl alcohol? (5) What is the purpose of using varying concentrations of isopropyl alcohol? How does this relate to the molecular structures of the dyes?

Extensions: Discuss health concerns of food dyes in the past. See *ChemCom* for an activity on the chromatography of food dyes in candy and Schneider et al. (2007) for a more advanced version of the *ChemCom* lab.

New problem: Most kidney stones are calcium stones (composed of calcium oxalate) and are formed according to the equation below:

$$Ca(HCO_3)_2 \text{ (aq)} + Na_2C_2O_4 \text{ (aq)} \rightarrow 2NaHCO_3 \text{(aq)} + CaC_2O_4 \text{ (s)}$$

Colas are high in oxalates and are not recommended for people who suffer from kidney stones. A can of cola contains (244 mg) of sodium oxalate, $Na_2C_2O_4$. How many cans of soda would someone have to drink to produce a kidney stone with a mass of 1.00 g (1000 mg), assuming that there is plenty of calcium hydrogen carbonate available in the body?

In *Living by Chemistry*'s toxins curriculum (Stacy et al., 2003c), students examine reaction types by looking at chemical changes that produce materials which, at certain concentrations, can be harmful to humans. Often, solution chemistry is introduced after chemical reactions; however, "Toxins" opens with an introduction to solutions and molarity to provide a context for the study of harmful chemicals. For students to track toxins, they must learn the language of chemistry (chemical equations). Later, students investigate precipitation reactions as the backdrop for stoichiometry and limiting reactants.

There are also some important biological reactions that involve smaller organic molecules that can be used to illustrate chemical reactions. One example is alcohol fermentation that is used by yeast and some bacteria ($C_6H_{12}O_6 \rightarrow 2C_2H_6O + 2CO_2$). As the products of this reaction are ethanol and carbon dioxide, humans have taken advantage of this reaction for years in the making of bread, beer, and wine. Students should easily be able to classify alcohol fermentation as a decomposition reaction.

Small organic molecules can be used in labs and stoichiometry problems as well. One good example that incorporates an organic acid with biological relevance is the making of fizzy drinks from citric acid and baking soda. In this activity, students work in the classroom or other location appropriate for food preparation (not the lab) and use stoichiometry to determine the correct proportion of ingredients (Rohrig, 2000). Students can easily evaluate their own results, as the incorrect ratio tastes bad. In addition to stoichiometry, this activity is ideal for equation writing and identifying the phases of reactants and products. Food-grade citric acid is inexpensive and may be purchased through online natural food outlets.

Thermodynamics

Thermochemistry is often taught after stoichiometry to help students understand how stoichiometric relationships dictate overall changes in energy during a chemical reaction. Some courses may delve deeper into thermodynamics later in the course. From qualitative descriptions of exothermic and endothermic reactions to quantitative descriptions of heat flow, entropy changes in reactions, and free energy changes used to predict reaction spontaneity, biological examples can help students make connections between biology and chemistry, as well as see the relevance of thermodynamics to their everyday lives. The most biologically significant process

---

Ready, Set, Break It Down Now

Purpose: Introduction to identifying chemical reactions

Background: Hydrogen peroxide (H_2O_2) is toxic to living organisms; therefore, it is quickly decomposed to yield H_2O and O_2 by the enzyme catalase. Although the decomposition of H_2O_2 is very slow under standard conditions; in one second, one molecule of catalase catalyzes the decomposition of 10^7 molecules of H_2O_2 (Moran et al., 1994).

Activity and assessment: Tell students that hydrogen peroxide (H_2O_2) is toxic to living organisms and ask them what compounds they think H_2O_2 could be broken down into that would be less toxic. If students say H_2 and O_2 gas, you can demonstrate how explosive this mixture is using a video clip from the internet (Purdue, 2002) or the *Journal of Chemistry Education* Chemistry Comes Alive video collection. Have students write a balanced chemical equation for the reaction they predict will occur. Add a small portion of fresh apple, fresh potato, or fresh raw liver to a test tube containing 5 ml of 3% H_2O_2 and describe what evidence indicates a chemical reaction is occurring (gas bubbles and test tube gets warm). You may want to have a brief discussion of catalysts at this point. You can also connect this activity to the elephant toothpaste demonstration (Spangler, 2007) and test the gas evolved with a glowing splint to confirm that it is oxygen.

Extensions: This can also be linked to kinetics and how catalysts or enzymes (nature's catalysts) speed up a reaction by lowering the activation energy. You can ask students what they might do to make the reaction go faster (for example, grind up the potato). Future labs can also include determining the optimal pH and temperature for enzymes to work. Example of such activities can be found in *ChemCom* (ACS, 2006).

that affects students is the metabolism of food. Metabolism can be examined on the macroscopic, cellular, and particulate levels. By showing students how chemical potential energy in foods can be converted into heat energy through combustion, they can appreciate the importance of understanding heat flow, calorimetry, and the relationship between heat and temperature in chemical changes. It is critical that experiments allow students to think about and discuss the difference between heat and temperature. Calorimetric experiments that allow students to calculate the energy released when food combusts can be used to help students relate joules to calories and kilocalories (Calories). Additionally, students can use the difference between energy released by food and energy gained by water (in a calorimeter) to explore what is meant by system, surroundings, and the First Law of Thermodynamics. Students should also come to understand that changes in energy do not always involve changes in temperature (e.g., phase changes). The standard exercise, in which students calculate the total number of calories required to change a known mass of ice at some temperature below 0°C to steam at some temperature above 100°C, could be made into a more interesting and biologically relevant problem by incorporating a weight loss myth. Because it takes energy to heat and/or melt cold water, ice, or even ice cream, some believe that drinking or eating these will lead to weight loss. The following problem helps students to distinguish between energy changes that involve temperature from those related to phase changes, highlights the difference between food Calories (kilocalories) and calories, and provides the "skinny" on this popular weight loss myth:

Suppose you eat two ice cubes (11.5 g) at −2.1 °C. Assuming that you don't choke or have a fever (your body temperature is 37.0°C), how much energy, in calories, is absorbed by the ice when it warms, melts, and warms to your body temperature? How many dietary Calories (kcal) were absorbed by the water? Is swallowing ice an effective weight loss method? Why or why not? Describe how the kinetic and potential energy of the ice change after you swallow it.

In addition to focusing on food to introduce the concept of calorimetry at the macroscopic level, revisiting the topic of cellular respiration from biology can link chemistry and biology, as well as focus on the particulate level, while addressing a common chemical misconception. Many students hold the belief that bond breaking is an exothermic process. A quick Internet search of high school biology Web sites reveals many statements describing how removing the third phosphate group of ATP releases a great deal of energy or free energy. The sites go on to define free energy as energy stored in chemical bonds. In addition, text and Web diagrams that depict the cleaving of the bond of one phosphate group from an ATP molecule often include an orange graphic labeled, "energy" appearing to be shooting out of the hydrolysis site. Although high school biology and chemistry courses are

The Heat Is On

Purpose: To determine the heat content of hydrocarbons using calorimetry

Background: Burning of hydrocarbons in air (O_2) produces H_2O and CO_2. These products are lower in energy than the original hydrocarbons; thus energy is released in the form of heat. Similarly, our bodies break down food, releasing energy, which is reported as dietary Calories. In chemistry, we define a calorie as the energy needed to raise the temperature of 1 g of H_2O by 1°C. The dietary Calorie is equal to 1000 of these calories.

Activity: Students will use either candles or small oil lamps to determine the heat content of hydrocarbon fuels. To do this, students will need to know the initial and final masses of the heat source and the mass of the water, as well as the initial and final temperature of the water. There are several variations on the calorimeter; a popular model uses a soda can with a stirring rod threaded through the pop top suspended from a ring on a stand (ACS, 2006). Allowing the water temperature to rise 20°C will provide good results. Students should use their data and the heat capacity of H_2O to calculate the heat content of their fuel source. An alternative to using hydrocarbon fuels is to use a food source such as a nut, a marshmallow, or a cheeto. (Note that some schools place restrictions on the use of food because of students' allergic reactions.)

Assessment: Students can be asked the following questions: (1) Which fuel source has the highest heat content? Rank the remaining fuel sources from highest to lowest. (2) Write the balanced chemical equation for the combustion of wax $(C_{25}H_{52})$. (3) What mass of oil is needed to raise the temperature of 2 L of water from the room temperature to water's boiling temperature? (4) One Krispy Kreme original glazed doughnut contains 210 dietary Calories of chemical energy; how much wax (or oil) would you have to burn to release the same amount of energy as the doughnut?

Alternative: Students can burn a food source, determine the amount of energy released, and compare this to the nutritional information on the package. Cheetos and marshmallows are good alternatives to cashews for those concerned about nut allergies (*Journal of Chemical Education* Editorial Staff, 2004; Stacy et al., 2003a).

David Amer, USAFA

responsible for the persistence of this misconception, this can be alleviated by drawing students' attention to the confusion surrounding this topic (Galley, 2004). A critique of resources showing the conversion of ATP to ADP may interest students and emphasize the notion that bond breaking is endothermic and bond making is exothermic. This could serve as an introduction to Hess's law, along with the sign conventions associated with enthalpies of formation (ΔH_f). As an extension, students may be interested in exploring macromolecules (fats, proteins, and carbohydrates) from a thermodynamic perspective. By using enthalpies of formation, students can determine why fat metabolism releases more calories of energy than the metabolism of proteins and carbohydrates.

Lastly, the Second Law of Thermodynamics can be introduced, but as it pertains to biological processes, such as photosynthesis and cellular respiration. Even a qualitative discussion of entropy is useful; however, be cautious, as defining a closed system around the complex cycle of photosynthesis and cellular respiration may be difficult. Students can summarize the net changes in enthalpy and entropy for photosynthesis and cellular respiration, with an emphasis on the energy changes that occur as a result of bond breaking and bond making and on the relative changes to the entropy of the system.

By recognizing where the physical and life science standards relevant to chemistry overlap, teachers can plan their courses to take advantage of students' experience and interest in the life sciences, thereby making chemistry accessible to more students.

Recommended Readings

ACS. *Chemistry in the Community*, 5th ed.; W. H. Freeman: New York, 2006. A thematic based ACS textbook that includes units on Food; Energy Storage and Use; and Proteins, Enzymes, and Chemistry. Each chapter also contains various activities, some of which are highlighted in this chapter.

Stacy, A.M.; Coonrod, J.; Claesgens, J. *Living by Chemistry: General Chemistry. Unit 5 Fire. Teacher's Guide,* preliminary ed., Key Curriculum Press: Emeryville, CA, 2003a.

Stacy, A.M.; Coonrod, J.; Claesgens, J. *Living By Chemistry: General Chemistry. Unit 2 Smells. Teacher's Guide,* Key Curriculum Press: Emeryville, CA, 2003b.

Stacy, A.M.; Coonrod, J.; Claesgens, J. *Living By Chemistry: General Chemistry, Unit 4 Toxins. Teacher's Guide,* Key Curriculum Press: Emeryville, CA, 2003c. A full year high school chemistry curriculum that is divided into 6 thematic units. Each unit consists of 25–30 guided inquiry lessons, each 50 minutes in duration. The student and teacher materials have been field tested in high school classrooms. Biological chemistry examples from these materials have been highlighted throughout this chapter.

Recommended Web Sites

ACS Chem Club Student Resources. http://portal.acs.org/portal/acs/corg/content?_ nfpb=true&_pageLabel=PP_ARTICLEMAIN&node_id=1505&use_sec=false (accessed March 11, 2008). An ACS Web site that has a number of links of interest to high school students. Under "What's that stuff" students can learn about materials such as lipstick, margarine, artificial sweeteners, food coloring, and more. There are also links for AAAS science podcasts, careers in chemistry, and reading up on chemistry topics.

McGill University Department of Chemistry. Office for Science and Society. http://oss. mcgill.ca/ (accessed March 11, 2008). A Web site highlighting chemistry in our everyday lives through its science issues and science in the news features. Additional features include a question-and-answer section and online lectures covering topics such as food chemistry, the diversity of chemistry, drugs, and chemistry and the environment. The lectures include visuals and audio.

References

American Chemical Society. *Chemistry in the Community*, 5th ed.; W. H. Freeman: New York, 2006.

Bodner, G. M. I have found you an argument: The conceptual knowledge of beginning chemistry graduate students. *J. Chem. Educ.* 1991, *68,* 385–389.

Bransford, J. D.; Brown, A. L.; Cocking, R. R., Eds. *How People Learn: Brain, Mind, Experience, and School,* National Academies Press: Washington, DC, 2000.

Emsley, J. Artificial Sweeteners. *ChemMatters*. February 1988, 4–8.

Galley, W. C. Exothermic bond breaking: A persistent misconception. *J. Chem. Educ.* 2004, *81,* 523–525.

Institute of Food Technologies. Food Chemistry Experiments, 2007. http://www. accessexcellence.org/AE/AEPC/IFT (accessed March 11, 2008).

Journal of Chemical Education Editorial Staff. Calories—Who's Counting? *J. Chem. Educ.* 2004, *81,* 1440A–1440B.

Kimbrough, D. R. The Aspirin Effect: Pain Relief and More. *ChemMatters*. February 2004, 7–9.

Lyndaker, A. M. The Molecules Around Us. Cornell University. Graduate Student School Outreach Project Web site, 2006. http://www.psc.cornell.edu/gssop/courses/ Molecules_Around_Us/2006/index.php?page=11 (accessed March 11, 2008).

Medline Plus Medical Encyclopedia. CHEM-20, 2005. http://www.nlm.nih.gov/medlineplus/ ency/article/003468.htm (accessed March 11, 2008).

Moran, L. A.; Scrimgeour, K. G.; Horton, H. R.; Ochs, R. S.; Rawn, J. D. *Biochemistry,* 2nd ed. Prentice Hall: Englewood Cliffs, NJ, 1994.

Ontario Ministry of Agriculture, Food, and Rural Affairs. Food Ingredients, 2007. http://www.omafra.gov.on.ca/english/food/industry/food_proc_guide_html/ chapter_4.htm (accessed March 11, 2008).

Ophardt, C. E. *Virtual Chembook,* 2003. http://www.elmhurst.edu/~chm/ vchembook/582dnadoublehelix.html (accessed March 11, 2008).

Purdue University. The Chemistry of Hydrogen, 2002. http://chemed.chem.purdue.edu/ demos/main_pages/10.1.html (accessed March 11, 2008).

Purves, W. K. Ask the Expert: How can an artificial sweetener contain no calories?, November 27, 2006. Scientific American.com web site. http://www.sciam.com/ askexpert_question.cfm?articleID=0007F523-93FA-1CE2- 93F6809EC5880000&catID=3&topicID=4 (accessed March 11, 2008).

Rohrig, B. Fizzy drinks: Stoichiometry you can taste. *J. Chem. Educ.* 2000, *77,* 1608A–1608B.

Schneider, R. F; Kerber, R. C.; Akhtar, M. J. Identification of Food Dyes, 2007. http://www.ic.sunysb.edu/Class/che133/susb/susb009.pdf. (accessed March 11, 2008).

Spangler, S. Steve Spangler Science, 2007. http://www.stevespanglerscience.com/ experiment/elephants-toothpaste (accessed March 11, 2008).

Stacy, A.M.; Coonrod, J.; Claesgens, J. *Living by Chemistry: General Chemistry. Unit 5 Fire. Teachers Guide,* preliminary ed., Key Curriculum Press: Emeryville, CA, 2003*a*.

Stacy, A.M.; Coonrod, J.; Claesgens, J. *Living by Chemistry General Chemistry: Unit 2 Smells. Teachers Guide,* Key Curriculum Press: Emeryville, CA, 2003*b*.

Stacy, A.M.; Coonrod, J.; Claesgens, J. *Living by Chemistry General Chemistry: Unit 4 Toxins. Teachers Guide,* Key Curriculum Press: Emeryville, CA, 2003*c*.

University of Minnesota Chemistry Department Lecture Demonstration Services. Separation of Food Dyes via Column Chromatography, 2000. http://www.chem.umn.edu/services/lecturedemo/info/column_chromatography.html (accessed March 11, 2008).

Chapter 6

Earth System Science Topics in the News: How Teachers Can Use These Contexts To Teach Chemistry and Inquiry

by Ann Benbow and Cheryl Mosier

Ann Benbow is the director of education, outreach and development at the American Geological Institute, where she manages curriculum development, teacher education programs, outreach to the general public, and research into the status of Earth system science education in the United States and beyond. A graduate of the University of Maryland in College Park, she has taught science to students in grades K–16. Contact e-mail: aeb@agiweb.org

Cheryl A. Mosier is an earth science teacher at Columbine High School in Littleton, CO. She also works as an educational consultant for It's About Time Publishing (a division of Herff Jones) and for the American Geological Institute. Contact e-mail: camosier@jeffco.k12.co.us

Introduction

The earth sciences provide many natural links to chemistry topics and are an excellent way of introducing chemistry to students early in their high school careers. Using an earth system science approach, students can investigate the chemical aspects of the *hydrosphere* (composition, water cycle, water quality, role as a solvent), the *atmosphere* (composition, air quality, acid rain), *geosphere* (soil chemistry, composition of volcanic eruptions, composition of rocks and minerals, chemical weathering), and the *biosphere* (biochemistry, carbon cycle, effects of pollutants on the biosphere).

This chapter describes ways in which teachers can use earth science topics that are of great interest to students and are frequently in the news to serve as contexts for chemical investigations. The topics in this chapter include "earth as a connected set of systems," "earth change over time," "natural hazards," and "water and the earth system." Suggestions for activities provide opportunities for students to not only progress toward proficiency in the science as inquiry standards, but also to make connections between the chemistry-related physical science standards and the earth and space science standards (see Table 1).

Mike Ciesielski

Table 1. National Science Education Standards for physical science and earth and space sciences

Content Standard B: Physical Science	Content Standard D: Earth and Space Science
• Structure of atoms • Structure and properties of matter • Chemical reactions • Motions and forces • Conservation of energy and increase in disorder • Interactions of energy and matter	• Energy in the earth system • Geochemical cycles • Origin and evolution of the earth system • Origin and evolution of the universe

Reprinted with permission from National Science Education Standards © 1995 by the National Academy of Science, courtesy of the National Academies Press, Washington, D.C.

Earth as a Connected Set of Systems

Until recently, the earth sciences tended to be studied and taught as independent subjects. Now there is a strong movement toward studying the Earth as a set of *interconnected systems*: the **hydrosphere** (water), **geosphere** (solid part of the planet), **atmosphere** (air), and **biosphere** (all life on the Earth). One of the ways in which these Earth systems are studied today is through remote sensing. This involves the use of satellite technology to gather data on cloud cover, aerosol movement, ozone concentrations, gravity variations on the planet, water temperature, chlorophyll production, storm tracking, and much more. Using data from remote sensing satellites is an engaging and unusual context for introducing such chemistry topics as the formation of various types of aerosols and their potential effects on the environment, ozone formation around the planet, and the role of ozone in the atmosphere.

Suggested Activities

Teachers can introduce students to a number of chemistry concepts using online satellite data. The Aura satellite, for example, monitors ozone concentrations. Students studying the formation and effects of ozone on the planet can access these data and build their own animations showing changes in global ozone concentrations over set periods of time. They can then hypothesize about factors contributing to the changes in these levels. Students interested in biochemistry can explore the role of chlorophyll in the carbon cycle by using data collected by the Sea-viewing Wide Field-of-view Sensor (SeaWiFS) instrument on the SeaStar satellite. The data from both Aura and SeaWIFS are on NASA's Earth Observatory Web site (see Recommended Web Sites at the end of this chapter).

Additional questions that students could investigate include:
- In which directions do aerosols move around the globe?
- What types of aerosols end up in the atmosphere and how are they formed? What are their chemical compositions?
- How does ozone form and how are satellites able to measure its concentration?
- Why is it important to keep track of ozone concentrations?

Earth Change Over Time

Scientists use a variety of evidence to track how the Earth has changed over its 4.5 billion year life span. Fossil evidence collected by paleontologists is extremely important in putting together the picture of how life forms developed, were related, spread, altered, and became extinct. Paleoclimatologists, on the other hand, are interested in how Earth's climate has changed over time. One of the data sets they collect is the relative levels of oxygen isotopes in ice cores (related to *structure of atoms*). These levels are used to determine when ice covered different parts of the earth and when those ice sheets receded. Using the ice core evidence as a context, students can be introduced to the concept of isotopes and how they are used in many ways as "trackers" in scientific research. This same context could be used to introduce students to the properties of gases.

Suggested Activities

Students could degas samples of carbonated beverages through heating. They could capture the released gas and test it by passing it through a solution of bromthymol blue (the carbonic acid formed by the carbon dioxide in water will turn the solution yellow). Students can also graph the changes in carbon dioxide, methane, or oxygen isotopes over time using data taken from ice cores in Antarctica. The National Oceanographic and Atmospheric Administration Web site http://www.ncdc.noaa.gov/paleo/icecore/antarctica/antarctica.html contains data from a variety of locations in Antarctica.

The curriculum *Earth System Science in the Community: EarthComm* (AGI, 2001) includes an activity in the *Earth System Evolution* unit that is very similar to the above online activity. In the chapter on climate change, the activity entitled *How Do Carbon Dioxide Concentrations in the Atmosphere Affect Global Climate?* has students work through a graphing activity (CO_2 vs. temperature) and interpret a graph (CH_4, CO_2, and temperature variations), looking for correlations among the three variables. The activity section ends with students creating their own experiments that demonstrate the greenhouse effect in the atmosphere. Students then read a short section called "Digging Deeper" that discusses CO_2 and global climate, greenhouse gases, and the carbon cycle.

Additional questions that students could investigate include
- How do chemists identify different gases?
- How do gases get trapped in liquids? How is it possible to get gases out of liquids?
- How are the isotopes of oxygen different from one another? How can these be used in paleoclimatological research?
- How are isotopes of various elements used as "trackers" in scientific research?

Natural Hazards

Natural hazards are of great interest to secondary school students. Hurricanes, earthquakes, tornadoes, tsunamis, volcanic eruptions, landslides, blizzards, dust storms, sink holes, and floods happen all over the world, every day. The hazards that link most readily to introducing chemistry concepts are volcanic eruptions, floods, and tsunamis. Any hazard that disrupts the water supply can work as a method for introducing solubility, units of concentration, drinking water purification processes, and wastewater treatment. Volcanoes are an interesting context in which to explore the chemical composition of matter erupting from volcanoes and its effects on the environment.

The eruption of a volcano puts particles into the atmosphere that can result in acid rain (related to *chemical reactions*). In fact, isotopes of sulfur from volcanic eruptions are being used to study climate change. A team of American and French scientists recently reported that they were able to study the effect of major volcanic eruptions on climate change by measuring the sulfur isotopic "fingerprint" of several relatively modern major volcanic

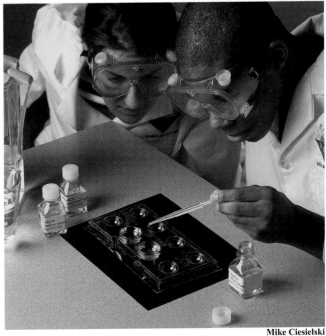

Mike Ciesielski

eruptions. Those that shot particles into the stratosphere (where the particles were able to reflect enough sunlight to lower the Earth's temperature) had different concentrations of sulfur isotopes than volcanoes with weaker eruptions (related to *structure and properties of matter*).

Understanding the chemical structure of magma offers students an opportunity to understand how viscosity has an impact on lava flows. The viscosity of lava controls the shape of a volcano and determines whether the volcano will erupt explosively like Mt. St. Helens, or ooze like Kilauea. Lava viscosity is determined mostly by the silica content within the lava, along with the temperature of the magma. Knowing the type of lava that may come out of a volcano helps geoscientists determine where and when people should be evacuated.

Suggested Activities

Students could investigate the pH of simulated volcanic products. Students could also create and test the pH level of their own acid lake using carbon dioxide bubbled through water. They could then use their acid lakes to investigate the rate of chemical weathering on small samples of limestone, marble, and other common rocks.

EarthComm, Earth System Science in the Community has an investigation on viscosity in the "Earth's Dynamic Geosphere" unit. During the investigation "Volcanic Hazards: Flows," students model the viscosity of lava using liquid soap. Students first investigate how much area is covered using increasing volumes of soap, and then create their own investigation, changing the slope of the land, viscosity of the soap, or using channels. Most students compare temperature effects on the soap, which models the change in viscosity based on temperature of the lava. Some also choose to add something to the soap to thicken it, modeling the effect that more silica has on lava. Students then complete a graph that shows the travel times of lahars (volcanic mudflows) from Mt. St. Helens and read about lava, pyroclastic flows, and lahars.

Sample questions students could investigate include
- How does acid rain form from volcanic products and how acidic is this rain?
- How does acid rain affect limestone and marble?
- How can gases released from volcanoes have an impact on groundwater?
- How does the viscosity of a liquid affect how fast a liquid flows?

Water and the Earth System

Water is a topic constantly in the news. Whether droughts in Ethiopia, floods following a hurricane in Louisiana, or leaking sewers in Rio de Janeiro, water's effect on the planet is profound. Investigations centering on water issues afford rich opportunities for students to explore the chemistry standards through Earth science examples. Students can begin by studying the structure and composition of the water molecule for clues as to why it is such an important substance on the planet (related to *structure and properties of matter*). They can then move on to the properties of water that, in combination with other substances and energy, contribute to weather, erosion, ocean currents, the water table, watersheds, water pollution, climate, and much more (related to *interactions of energy and matter*). Teachers could also create a series of lessons on the role of chemical reactions in monitoring and cleaning water (related to *chemical reactions*).

Suggested Activities

People who live close to large bodies of saltwater are frustrated with the lack of drinking water available to them. Having students investigate the differences between freshwater and saltwater helps them to understand why ocean water is not usable by humans. This also helps them realize why desalination is so expensive and not an easy option for many communities.

When students compare the properties of fresh- and saltwater through a variety of activities, they gain a better understanding of the chemistry of water and the characteristic properties of both salt-water and freshwater.

One activity involves students calculating the density of freshwater vs. saltwater. Students measure the mass and volume of a sample of each type, calculate their densities, and then compare them. After the density is calculated, students can then predict whether objects of known density will sink or float in each type of water. Predictions can be followed by testing. This can also be done as a demonstration with eggs, which will float in saltwater, but not in freshwater. Next, students can compare the boiling points of both fresh- and saltwater. Students collect and graph temperature data on both types of water to discover that saltwater boils at a higher temperature than freshwater.

Students can simply evaporate both types of water in a shallow pan or clean Petri dish to compare the "stuff" left behind, or they can model the process of distillation. To model distillation, you need a small clear dish (heavy enough to sink in water), a large clear bowl, plastic wrap, some type of small weights, saltwater, and either a sunny location or an overhead projector. A small amount of saltwater is placed into the large bowl, about 5 cm deep, but dependent on the size of the small bowl. The small bowl is placed into the large bowl, carefully keeping the small bowl dry on the inside. The large bowl is covered and sealed with plastic wrap, snugly, but not tightly. The small weights are placed in the middle of the plastic wrap directly over the small bowl, and then the whole setup is carefully moved to a sunny location or placed on a running overhead projector. Over time, the water will evaporate, condense on the inside of the plastic wrap, and then drip into the small bowl. The amount of time this takes will depend on the ambient temperature of the room and the amount of heat supplied by the sunshine or overhead projector.

These activities allow students to investigate the properties of both fresh- and saltwater. They also give them an understanding to support further studies of water purification, desalination, and oceanography.

Sample questions that students could investigate on their own or in small groups can include
- What is the role of water in shaping the Earth?
- Why is water important in the weather machine?
- How does water become polluted?
- How can polluted water be cleaned?
- Why is ocean water salty?
- Why can't living things survive drinking saltwater?
- What is the role of water in photosynthesis? In respiration?
- How does water get into the ground?
- How does desalination work?
- How is flood water able to cause so much damage?

Recommended Web Sites

http://www.earthobservatory.nasa.gov This Web site, developed by the Terra satellite team at NASA, contains satellite data, animations, activities, news, and a glossary of Earth system science terms. Students and teachers can create their own animations on this Web site using remote sensing data from a number of the NASA satellites. (accessed March 12, 2008).

http://www.agiweb.org/earthcomm/ The American Geological Institute's Education Department maintains a Web site of state-based resources to teach high school earth science. These resources include activities, maps, geological data, and links to state geological surveys. (accessed March 12, 2008).

Recommended Readings

American Geological Institute. *Earth System Science in the Community: EarthComm*. It's About Time Publishing: Armonk, NY, 2001: This five-module curriculum is based on the earth science standards of the National Science Education Standards and was funded by the National Science Foundation. It is appropriate for grades 9–12.

American Geological Institute. *Investigating Earth Systems*. It's About Time Publishing: Armonk, NY, 2002: This ten-module inquiry-focused curriculum is based on the earth science standards of the National Science Education Standards and was funded by the National Science Foundation. It is appropriate for students in grades 6–9.

References

American Geological Institute. *Earth System Science in the Community: EarthComm*. It's About Time: Armonk, NY, 2002.

National Research Council. *National Science Education Standards*. National Academies Press: Washington, DC, 1996.

Chapter 7

Technology Standards and the Chemistry Laboratory

by Loretta Jones and Seán P. Madden

Loretta Jones *(Ph.D. and D.A., University of Illinois at Chicago) is professor of chemistry at the University of Northern Colorado and was the 2006 Chair of the Chemical Education Division of the American Chemical Society. Her research area is in chemical education, particularly, the active involvement of students in their learning and the applications of advanced technologies. She is a principal developer of award-winning multimedia chemistry courseware, and she has led a large high school chemistry curriculum development project.*
Contact e-mail: Loretta.Jones@unco.edu

Seán P. Madden *obtained both his Ph.D. in chemical education and M.A. in chemistry from the University of Northern Colorado, a B.S. in nuclear technology through the U.S. Navy and Excelsior College, and a B.A. in molecular biology and science education from Colorado University, Boulder. He has taught high school science and mathematics for seven years and is currently employed at Greeley West High School in Greeley, CO. Contact e-mail: sndmadden1@juno.com*

If we were to imagine an ideal high school chemistry classroom, we might envision students doing inquiry-based, hands-on activities and using technology to enhance their learning. Yet, in actual classrooms, laboratory activities can be time consuming, expensive, and are sometimes overlooked, despite their importance. The standards for learning about technology and for laboratory instruction are broad and allow for a variety of interpretations. In general, inquiry experiences, many of which involve hands-on work with chemicals, are recommended (see chapter 4). However, even simple investigations can inspire and enlighten students. Questions such as how much time should be allotted for laboratory work, what a properly equipped laboratory should contain, and what types of laboratory activities are the most important, are left open to interpretation by individual teachers and school districts. In this chapter, we summarize current thinking about how to provide students with good laboratory experiences and share a variety of ways in which teachers can enrich classroom instruction with technology.

Laboratory Learning

The high school chemistry teacher is typically faced with limited resources and time; it is difficult under these conditions to conduct an exemplary laboratory program. Fortunately, help is available. The American Chemical Society (ACS, 2003) has produced a booklet on chemistry teacher preparation that also includes guidelines for managing laboratory work. In addition, the National Science Teachers Association (NSTA) has developed a set of guidelines for ideal high school science laboratories (Biehle et al., 1999). However, simple, low-cost and small-scale equipment, which has the advantage of reducing hazards and the amount of waste produced, can be used to introduce students to scientific inquiry and basic laboratory skills in nearly any classroom (Waterman and Thompson, 2000; Towse and Huseth, 1997; also see http://ssc.mriresearch.org).

In any kind of hands-on activity, safety is a primary concern. The ACS booklet on teacher preparation provides basic information on producing a safe environment in the high school chemistry laboratory (Tinnesand, 2007). Another useful resource for high school chemistry teachers is the Flinn Scientific catalog, which provides comprehensive information on how to store chemicals safely (Flinn, 2007). In addition, the NSTA has published a safety handbook for high school teachers that provides guidelines for working safely with chemicals (Texley, 2004). Green chemistry activities also offer options to improve safety and reduce waste, given their emphasis on the use of nontoxic, environmentally friendly methods, and chemicals (La Merrill et al., 2003).

Table 1. Science and Technology Standards

Grades 9–12
Abilities of technological design Understanding about science and technology

Source: NRC, 1996, p. 107.

Table 2. Changing emphasis on technology and laboratory as a result of NSES

Less of this	More of this
Purely paper-and-pencil or drill-and-practice exercises for numerical problem solving	Regular use of computers and graphing calculators to enhance quantitative understanding of chemistry
Spending an entire laboratory session simply setting up equipment and gathering data	Regular use of computer and graphing calculator interfaces with data collecting instruments that allow simultaneous data analysis and interpretation
Viewing data collection and graph making as an end in itself	Discussing the meaning of data and graphs, and connecting them to molecular phenomena
Use of outdated and unsafe practices that may lead to injury	Consistent use of appropriate safety procedures in all laboratory and demonstration settings
Conveying the behavior of chemical systems at the molecular level only with words	Using animations and molecular modeling software to facilitate visualization of molecular-level phenomena
Learning and teaching only with text	Multimedia animations to enhance conceptual understanding of chemistry and to provide additional inquiry experiences

Laboratory and Technology in the National Science Education Standards

The *National Science Education Standards* (NSES) for science and technology are very brief (Table 1), yet employing technology in the classroom and laboratory can significantly enhance student learning (Table 2). When computers or graphing calculators are used in the laboratory, data can be interpreted immediately and may therefore be more meaningful to students. Computers also allow teachers to introduce their students to molecular visualizations so that they can build more accurate mental models of the particulate level of matter (Kelly and Jones, 2007).

Technology in the Laboratory

Microcomputer-interfaced laboratory experiments make the introductory chemistry laboratory a new experience for students (and teachers, too!). Instead of repeating the same experiments that were completed by students a generation ago, today's students can have access to equipment that will collect data in a shorter period of time and present it in a more meaningful format. The emphasis of the laboratory experience can then shift to learning the concepts underlying the measurements, rather than on tedious weekly repetition.

Computer technology affords students of the 21st century a powerful opportunity to understand the intimate connection between science and mathematics. *Mathematical* models, (executed by computers) complement *chemical* models (which seek to bridge our macroscopic observations with their underlying microscopic, molecular basis). Inexpensive graphing calculators and the data collection devices that interface with them are the ideal computer technology for use in K–12 school setting.

Graphing calculators, such as those manufactured by Texas Instruments, Casio, Hewlett Packard, and Sharp, have the following capabilities:

Figure 1. Titration Experiment: A Traditional Approach

Suppose a group of students were titrating 50.00 ml of a weak acid (0.1000 M acetic acid), with 0.1000M NaOH. Without access to graphing calculators or a computer, students would titrate to an endpoint and collect only one data point for the titration. Students can connect their data to a stoichiometric calculation but will not see how the pH changes during the titration.

- Spreadsheet features in which data can be stored, graphed in a variety of formats (such as scatterplots), and transformed into using built-in mathematical functions like multiplication by a constant, inverses, logarithms, etc.;
- Regression functions, such as polynomial, sine, and logistic equations to analyze data or scatterplots of these data;
- Equation editors in which students can build their own mathematical functions to model the data contained in the list feature. Equation editors allow a function to be graphed, traced, solved for roots/maxima/minima, integrated numerically, and to have derivatives determined at specific points along the curve;
- Matrix algebra operations that allow for multiple linear regression of a data set, simultaneous solution of systems of equations, and other statistical treatments of data beyond the built-in capabilities of a calculator;
- Interfaces with data collection instruments, such as those manufactured by Vernier (Texas Instruments and Casio lines of calculators); and
- Programmability, allowing the user to design software that, among other things, controls the sample rate of an instrument and the display format of the collected data.
- All of this computing power comes in the form of an inexpensive hand-held device. Thus, graphing calculators fulfill much of the promise of our computer age and seem also to be the ideal choice of computer technology for the K–12 classroom.

An activity commonly carried out in the high school chemistry classroom and demonstrating many of these points can be found in Fig. 1. One of the goals of such a laboratory activity is that students discover an appreciation for the connection between the macroscopic phenomenon of pH and the microscopic, molecular behavior of the chemical species involved in the equilibrium reaction. Another goal is that students will discover the power of a mathematical model, in this case, the equilibrium constant, to connect these macroscopic and microscopic realms. Fig. 2 describes how the graphing calculator can serve as a great teaching and learning tool in this situation.

This same theme of collecting data and generating mathematical models to explain the data can be applied throughout the chemistry curriculum. When graphing calculators and their associated data collection devices are made available in the classroom, many opportunities arise for exploring the connections between mathematics and science, which further illuminate the connection between macroscopic and microscopic dimensions of chemistry.

A variety of resources are available to teachers who want to incorporate graphing calculator and data-gathering technology into their chemistry classrooms. These resources can be found in publications such as the *Journal of Chemical Education* and from publishers of high school mathematics curricula and chemistry curricula. Vendors of these materials include Vernier, Texas Instruments, Casio and Ocean Optics, Inc. Vendors' philosophies about their own technology range from those who seek to provide teachers and students with painless, black box data gathering and analysis tools, to those who seek to teach chemistry and science through activities that encourage students and teachers to develop their own ideas, including writing their own software programs.

Technology for Conceptual Learning

Technology can be used to not only support and enhance laboratory instruction, but also to enhance the learning of chemistry concepts both in the classroom and during homework. Modern chemistry and biology focus on the structure and properties of molecules. Molecular visualization programs enable scientists to create and manipulate dynamic representations of molecular structures that are otherwise hard to visualize. Such software can radically change the introductory chemistry curriculum by allowing a much earlier, and more central, focus on molecular structure and properties. Visualization programs make abstract chemical concepts more real and meaningful to students. However, students must develop those visualization skills. Meaningful, independent use of these tools by students requires guidance. For example, the *Chemsense* program, which allows students to build and animate molecular structures, provides orientation information for teachers and guidance for students (SRI, 2005). The *ChemDiscovery* program, which is a comprehensive one-year chemistry course on computer, offers extensive support materials for teachers (Agapova et al., 2000).

Virtual laboratories that make use of multimedia software allow students to view chemical reactions too hazardous to view in person. They can also promote rapid transfer of learning to the actual situation and allow the experiments used in the laboratory to be upgraded. The combination of hands-on

Figure 2. Titrations Redux: Taking Advantage of Technology Approach

With a Vernier pH probe connected to the data collector of a graphing calculator, students would be able to collect data such as that in Table 3 (reproduced from Skoog et al., *Analytical Chemistry,* 7th edition, 1996, p. 212, a college text, which contains a table of data for this common high school activity). Students enter the data into the spreadsheet feature of the calculator as shown in Fig. 3. (Note: Although the screen shots shown here are from the CFX-Casio 9850 GB Plus, the displays of Texas Instruments, Hewlett Packard, and Sharp calculators are very similar). Data are displayed using a view window (Fig. 4), and as a scatterplot shown in Fig. 5. (By default, the graphing calculator leaves the axes unlabeled; however, x- and y-axis labels such as "pH" for the y-axis and "volume of titrant" for the x-axis can be easily added.)

The curved appearance of the data might come as a surprise or discrepant event to the novice high school student. A guided inquiry discussion (see chapter 4) can lead to the following form of the Henderson-Hasselbalch equation, which describes the equilibrium behavior of weak acid systems:

$$pH = pK_a + \log [A^-]/[HA].$$

This equation can then be modified and entered into the equation editor (Fig. 6) of the calculator:

$$Y1 = 4.7 + \log (X \div (50 - X)),$$

where Y1 represents pH over the course of the titration, pK_a = 4.7 for acetic acid, X represents the variable concentration of acetate ion during the titration, and 50 represents the analytical millimoles of acetic acid present per liter. Graphed on the same view window as the data, students may be pleasantly surprised to find a similarly shaped curve (Fig. 7).

When the theoretical curve is superimposed on the raw data, students will undoubtedly notice a satisfying agreement with the mathematical model (Fig. 8). They may want to perform "mathematical experiments" to find another section of curve that matches the data for the remainder of the titration. Or, they may want to trace along the curve with the derivative feature in order to gain a sense of how quickly pH changes at different times during the titration (Fig. 9).

and virtual labs makes it possible to *increase* the amount of skills training, without increasing the time spent in lab. For example, in one school, time constraints did not allow students to perform dilutions in lab using volumetric glassware nor to prepare their own calibration curves for instruments. When multimedia software was introduced, it became possible to develop new hands-on experiments that required students to make several dilutions and to construct a calibration curve (Jones and Smith, 1991).

The computer simulations in virtual laboratories make many more reactions accessible to students and permit repeated trials, certainly more than would be possible in a hands-on laboratory. Simulations permit students to work with systems not possible to include in the laboratory, such as explosive mixtures, toxic chemicals, and reactions carried out at remote sites. Furthermore, because the experimental observations can be immediately interpreted and analyzed with the aid of the computer program, content learning can be enhanced beyond what is possible with traditional methods. Good sources of reviewed and freely available computer-based instructional materials include Multimedia Educational Resources for Learning and Online Teaching (Merlot, 2007) and a free interactive online introductory chemistry textbook (Rogers et al., 2007). It is important to note that these simulations should be used to *enhance* instruction, not to replace it. The benefits of inquiry-based hands-on activities cannot all be replicated in a computer program.

Table 3. Data collected from a titration of 50.00 mL of 0.1000 M CH_3COOH, a weak acid, with 0.1000 M NaOH

Volume of NaOH (in mL)	pH
0.00	2.88
10.00	4.16
25.00	4.76
40.00	5.36
49.00	6.45
49.90	6.7.46
50.00	8.73
50.10	10.00
51.00	11.00
60.00	11.96
75.00	12.30

Source: Skoog et al., 1996, p 212. Reprinted with permission from *Fundamentals of Analytical Chemistry*,7E © 1996 Brooks/Cole, a part of Cengage Learning, Inc.

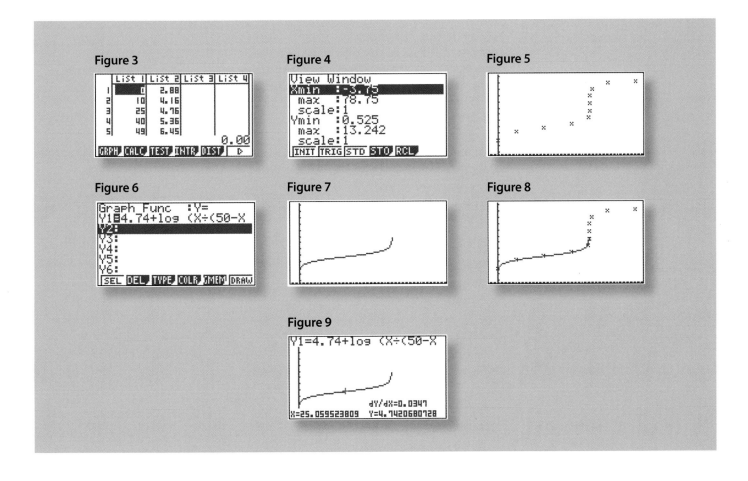

Figure 3

Figure 4

Figure 5

Figure 6

Figure 7

Figure 8

Figure 9

JupiterImages

Recommended Web Sites

Casio. http://casioeducation.com. (This Web site contains an archive of graphing calculator activities written by teachers that can be used in conjunction with all types of Casio graphing calculators) Merlot. http://chemistry.merlot.org (accessed July 13, 2007).

National Council of Teachers of Mathematics. http://www.nctm.org. The NCTM publishes several peer-reviewed journals. The *Mathematics Teacher* often contains articles demonstrating how graphing calculator technology and laboratory data can be incorporated into the classroom.

Ocean Optics. http://www.oceanoptics.com. This company manufactures and markets a number of data gathering instruments that can be interfaced with graphing calculators and computers.

SRI (2005). http://www.chemsense.org/ (accessed March 22, 2007).

Texas Instruments. http://www.ticares.com. (This Web site contains an archive of graphing calculator activities written by teachers that can be used in conjunction with all types of Texas Instruments graphing calculators).

References

American Chemical Society. *Safety in Academic Laboratories,* 7th ed. American Chemical Society: Washington, DC, 2003; Vol. 1 (student) and 2 (faculty). (One copy of each volume can be obtained free by calling 800-227-5558.)

Agapova, O. I.; Jones, L. L.; Ushakov, A. S.; Ratcliffe, A. E.; Varanka Martin, M. A. Encouraging Independent Chemistry Learning through Multimedia Design Experiences, Chem. Ed. Intl., 2007, 3, AN-8, 2002. Available at http://www.iupac.org/publications/cei/vol3/0301x0an8.html. Also see http://chemdiscovery.com/.

Biehle, J. T.; Motz, L. L.; West, S. S., NSTA Guide to School Science Facilities, NSTA Press: Washington, DC, 1999.

Carlson, R.J.; Winter, M.J. *Transforming Functions to Fit Data: Mathematical Explorations Using Probes, Electronic Data-Collection Devices, and Graphing Calculators*. Key Curriculum Press: Emeryville, CA, 1998.

Flinn. *Flinn Chemical and Biological Catalog Reference Manual*. Flinn Scientific: Batavia, IL, 2007.

Holmquist, D. D.; Randall, J.; Volz, D. L. Chemistry with CBL: Chemistry Experiments Using Vernier Sensors with TI Graphing Calculators and the CBL System. Vernier Software: Portland, OR, 1998.

Jones, L. L.; Smith, S. G. Using Interactive Video Courseware to Teach Laboratory Science, *Tech Trends* 1991, *35*, 22–24.

Kelly, R.; Jones, L. L. Exploring how different features of animations of sodium chloride dissolution affect students' explanations. *J. Sci. Ed. Technol.* 2007, *16*, 413–429.

La Merrill, M.; Parent, K.; Kirchhoff, M. Green Chemistry—Stopping Pollution Before It Starts. *ChemMatters* 2003, April, 7–10.

Madden, S. P.; Comstock, J.; Downing, J. P. Paper moon: Demonstrating a total solar eclipse. *Mathematics Teacher*. (December 2005/January 2006). (Please see annotated reference for the National Council of Teachers of Mathematics.)

Madden, S. P.; Rausch, L. Discovering Algebra: An Investigative Approach, Calculator Notes for the Casio fx-9750G Plus and CFX-9850GC Plus. Key Curriculum Press, Emeryville,

CA, 2007. (Contains programs for running experiments with equipment interfaced to the graphing calculator as well detailed instructions on how to use graphing calculators to analyze data).

Madden, S. P.; Runnels R. Discovering Algebra: An Investigative Approach, Calculator Notes for the Casio fx-7400G Plus. Key Curriculum Press, Emeryville, CA, 2007. (Contains programs for running experiments with equipment interfaced to the graphing calculator, as well detailed instructions on how to use graphing calculators to analyze data.)

Madden, S. P.; Wilson, W.; Dong, A.; Geiger, L.; Mecklin, Christopher J. Multiple linear regression using a graphing calculator: Applications in biochemistry and physical chemistry. *J. Chem. Educ.* 2004, *81,* 903. (This monthly publication of the Division of Chemical Education of the American Chemical Society regularly contains articles written by high school and college level chemistry instructors that illustrate how graphing calculators can be used in the chemistry classroom and laboratory.)

National Research Council. National Science Education Standards. National Academies Press: Washington, DC, 1996.

Rogers, E.; Stovall, I.; Jones, L.; Chabay, R.; Kean, E.; Smith, S. (2000). Fundamentals of Chemistry. Available at http://www.chem.uiuc.edu/webFunChem/GenChemTutorials.htm.

Skoog, D. A.; West, D. M.; Holler, F. J. *Fundamentals of Analytical Chemistry*, 7th ed., Saunders College Publishing: New York, 1996.

Texley, J.; Kwan, T.; Summers, J. *Investigating Safely: A Guide for High School Teachers.* NSTA Press: Arlington, VA, 2004.

Towse, P.; Huseth, A., Eds. *A Proceedings on Cost-Effective Chemistry: Ideas for Hands-On Activities.* Institute for Chemical Education: Madison, WI, 1997.

Waterman, E.; Thompson, S. *Addison-Wesley Small-Scale Chemistry Laboratory Manual*, 3rd ed., Prentice Hall: New York, 2000.

Chapter 8

Bringing Social and Personal Perspectives Into Standards-Based Chemistry Instruction in an Urban School District

by Donald J. Wink, Patrick L. Daubenmire, Sarah K. Brennan, and Stephanie A. Cunningham

Donald J. Wink *is a professor of chemistry at University of Illinois Chicago and director of graduate studies for a learning science program that integrates learning in the scientific disciplines with insight from other education-related fields. His work involves close collaboration with teachers and others in the improvement of teaching and learning in the Chicago Public Schools: the process of studying inquiry "close to the classroom," in authentic settings and on a large scale. Contact e-mail: dwink@uic.edu*

Patrick L. Daubenmire *earned a Ph.D. from The Catholic University of America in chemical education research. After ten years as a high school chemistry teacher and science department chairperson, Patrick joined the faculty at Loyola University Chicago, where he teaches general chemistry and serves as the Assistant Director of High School Programs in the Center for Science and Math Education. Patrick also manages the Inquiry to Build Content Science Instructional Development System, a jointly developed program of Loyola University Chicago and University of Illinois at Chicago within the Chicago Public Schools' High School Transformation Initiative. Contact e-mail: pdauben@luc.edu*

Sarah K. Brennan *has a B.S. in chemistry from DePaul University and a M.A.T. from Chicago State University. She taught chemistry and environmental science in the Chicago Public Schools for 6 years. Sarah currently works at the University of Illinois at Chicago, where she assists in the development of the Grade 10 chemistry curriculum for the Inquiry to Build Content Science Instructional Development System, a jointly developed program of Loyola University, Chicago and University of Illinois at Chicago within the Chicago Public Schools' High School Transformation Initiative. Contact e-mail: skbrennan21@gmail.com*

Stephanie A. Cunningham *has an M.S. in chemistry from the University of Illinois at Chicago and is currently pursuing a Ph.D. in Learning Sciences at the University of Illinois at Chicago. Twice she has received a National Science Foundation GK–12 Fellowship. Stephanie has been involved in the Inquiry to Build Content Science Instructional Development System, a jointly developed program of Loyola University, Chicago and the University of Illinois at Chicago within the Chicago Public Schools' High School Transformation Initiative. Contact e-mail: scunni2@uic.edu*

Introduction: Implementation of the National Science Education Standards' Standard F in Chemistry Curricula

Standard F in the *National Science Education Standards* (NSES) calls for instruction that "addresses ... the role of scientific knowledge in making decisions, relationships among science, technology, politics, and society, risk/benefit analysis, assessing alternatives, and the complexity of decision making" (National Research Council, 1996). This means that such knowledge has been placed on a par with that of scientific inquiry and traditional content knowledge. Besides such prescriptive notions for what students should know, there are also strong pedagogical and sociological reasons for teaching with a focus on connections between science and society. A recent report, *The Silent Epidemic* (Bridgeland et al., 2006), investigated why students dropped out from high school by polling these students. Almost half cited disinterest in classes as a reason why they dropped out, and this factor was cited more than any other, including poor performance. Additionally, when these students were asked what would have *kept* them in school, their most common response (cited by 81% of students) was improvement in "teaching and curricula to make school more relevant and engaging and enhance the connection between school and work."

The heart of Standard F and the ideas described in *The Silent Epidemic* suggest that relevant teaching will be more effective in helping students learn traditional content, yet relevance is a complicated entity to describe, let alone to create (Wink 2005). To be effective, curricula that use social and personal perspectives to frame learning must balance several factors not typically considered in traditional curricula. First, the integrity of students must be maintained so that the curriculum retains the necessary rigor, so students may learn the content. Second, students' *actual* interests should be identified over their *perceived* interests. These actual interests can be powerful tools for helping students notice science within their lives and environments. Third, gaps in students' background knowledge need to be addressed for both traditional content, as well as their social and personal perspectives: instructors cannot assume that students automatically know, for example, why air pollution affects the environment.

There are several different ways to address Standard F and the need to have students learn the way science knowledge affects society and their lives. Two will be considered here. On the one hand, instructors can attempt to insert relevance within existing curricula activities. This is versatile, but such efforts can be tangential and easily cut from a curriculum when content demands are high. Another alternative is to use curricula that embed the learning of traditional content in scenarios or themes that address social and personal perspectives. Several texts are now available to do this, including *Living by Chemistry* (Stacy, 2006), *Active Chemistry* (Freebury and Eisenkraft, 2006), and *Chemistry in the Community* (American Chemical Society, 2006).

In this chapter, we describe a multilevel, multiyear effort to use several science texts that have very strong emphases on Standard F of the NSES, in a district-wide initiative of Chicago Public Schools in order to increase the percentages of students who meet and exceed Illinois Learning Standards (Illinois State Board of Education, 1997). This initiative has the potential to be a model for other school districts, particularly those that face the challenge of high dropout rates and poor student preparation for high school, because these are at the center of the challenges this initiative addresses.

Our discussion of this initiative will focus on the chemistry portion of the curriculum, which uses *ChemCom*. In doing so, we will address five issues that arise when using a theme-based text to teach science and to attend to Standard F:

1. Articulating a national curriculum with a particular set of state standards.
2. Teaching in a spiral, need-to-know method and mapping content coverage.
3. Addressing inquiry, life science, and earth and environmental science standards while teaching chemistry from social and personal perspectives.

4. Embedding assessment that reinforces both content learning and learning about the social and personal perspectives of the curriculum.

5. Using the text (in our case, *ChemCom*) to assist teachers in addressing particular students' needs in the following areas:

    a. student motivation to learn,

    b. lack of preparedness from previous schooling experiences,

    c. appropriate instruction for English language learners (ELL) and students with special needs,

    d. development of both skills and thinking of "gifted students," and

    e. relevance of context-rich themes to students growing up in an urban environment.

Any instructor teaching with a theme-based textbook will encounter variations of the first four issues. Our examples, though, draw from our specific work with *ChemCom*. Therefore, the last issue relates to some specific opportunities while working with *ChemCom* in a context committed to building content understanding through inquiry teaching with a large number of urban students.

Articulating With State and National Standards

Any educator teaching in a K–12 public school must understand his or her state standards and associated assessment systems. We work in the specific context of Chicago through participation in the High School Transformation Project. The project assists schools who have low percentages of students who meet and exceed state learning standards. The first cohort of 14 schools formed in fall 2006. The second cohort includes 11 schools beginning in 2007, and a third cohort of up to 20 schools will join in 2008. Selected schools are provided with a three-year Instructional Development System (IDS) in each of mathematics, English, and science. Each IDS involves current and research-based curricula and pedagogy, teacher training and coaching, materials and resources, and clear, well-defined tools of assessment. For our project, a team from Loyola University Chicago and University of Illinois at Chicago provides an

Mike Ciesielski

"Inquiry to Build Content" science IDS. First-year students learn biology using *Biology: A Human Approach* (Biological Sciences Curriculum Studies, 2006), then proceed in the second year with *Chemistry in the Community* (American Chemical Society, 2006), and finish the IDS in the third year with *Active Physics* (Eisenkraft, 2004).[1] In order to ensure appropriate and meaningful implementation of our assessment program, professional development activities help coaches and teachers understand the benefits of data-driven decision making. Teachers also have ample opportunity and support to use assessment data to make instructional decisions at both a fine-grained and a "big-picture" level.

In our case, we needed our curriculum to work with both the NSES and the Illinois Learning Standards (ILS). Of course, some of this had already been done, at the national level, when the developers of *ChemCom* created a detailed description of how it aligns with the content standards of the NSES (Fig. 1).

However, because the NSES are only advisory, the "real" standards for us were the ILS. These form the basis of a set of statewide assessments for public school students at the end

---

[1] Concurrent with or following completion of this IDS, students may choose to take science electives at their respective schools. These courses were not included as a necessary part of the science IDS within the transformation initiative.

Figure 1. Articulation of ChemCom with NSES Content Standards

Overview: NSES Content Standards, Grades 9–12, vs. *ChemCom* Coverage						
A. Science as Inquiry	B. Physical Science	C. Life Science	D. Earth and Space Science	E. Science and Technology	F. Science in Personal and Social Perspectives	G. History and Nature of Science
Abilities necessary to do scientific inquiry	Structure and properties of matter	The cell	Energy in the earth system	Abilties of technological design	Personal and community health	Science as a human endeavor
Understandings about scientific Inquiry	Structure and properties of matter	Molecular basis of heredity	Geochemical cycles	Understandings about science and technogoly	Population growth	Nature of science knowledge
	Chemical reactions	Biological evolution	Origin and evolution of the earth system		Natural resources	Historical perspectives
	Motions and forces	Interdependence of organisms	Origin and evolution of the universe		Environmental quality	
	Conservation of energy and increase in disorder	Matter, energy, and organization in living systems			Natural and human-induced hazards	
	Interaction of energy and matter	Behavior of organisms			Science and technology in local, national, and global challenges	

(Left vertical axis label: Fundamental abilities, concepts, principles)

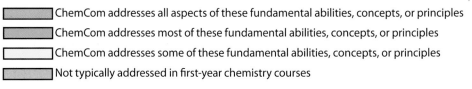

ChemCom addresses all aspects of these fundamental abilities, concepts, or principles

ChemCom addresses most of these fundamental abilities, concepts, or principles

ChemCom addresses some of these fundamental abilities, concepts, or principles

Not typically addressed in first-year chemistry courses

of the junior year known as the *Prairie State Achievement Examination* (PSAE). The PSAE includes the ACT exam and also content-focused subject tests in English, mathematics, and science. The subject tests have been written by the State of Illinois following a series of framework statements linked to the ILS. These framework statements do indeed include statements related to social and personal perspectives, meaning that for students to do well they must be able to link their knowledge to socially and personally relevant questions.

We showed how our curriculum met these standards in a "bottom-up" fashion. We took the most specific standards (the PSAE framework statements) and determined where they are covered in our curriculum. We also noted that certain things are, inexplicably, missing (these include redox reactions, for example), so we mapped the additional content that we, as chemists, knew our students needed to prepare for college. Because the PSAE framework statements had already been mapped to the ILS and the links between the ILS and NSES had already been shown, we therefore had a standards map for our program.

However, in our program, we had another requirement, because we are actually working not just with 10th grade chemistry, but also with biology and physics. This created both a need and an opportunity to create a single, comprehensive, three-year vertical alignment of scientific concepts and skills (Table 1). By making connections to these overarching themes (see chapter 3) among the three independent curricula, we could support teaching and learning about a more comprehensive understanding of science. These themes have the promise that knowledge will build to an authentic perspective that the student holds, formed alongside content knowledge because the content is learned from a social and personal perspective consistently and over three whole years.

Table 1. Conceptual Goals of Science Instruction

Conceptual Goal Standards Alignment	Science Skill Set
Matter and energy	Design and conduct investigations
Evolution	Use mathematical reasoning
Interactions	Using evidence to make informed decisions
Force and motion	Interacting with groups
Inquiry and technology	Communicating scientifically
Nature of science	

An example of the advantage of this vertical alignment comes from lessons coded as addressing "matter and energy." This allows us to explicitly link instruction in biology (metabolism and personal nutrition) to chemistry (atmospheric chemistry and the environment) and to physics (electrical energy and the home). Note that in all three cases, the vertical alignment highlights how a single scientific theme occurs in three different sciences *and* in three (or more) social and personal perspectives. Thus, social and personal perspectives become the basis for reinforcing the unity of science across multiple standards.

Teaching on a Need-to-Know Basis

As mentioned, teaching from a theme-based text creates excellent opportunities to address Standard F, as Fig. 1 shows for *ChemCom*. However, there is still the important question of how these texts address the teaching of traditional chemistry content. There is a tension that arises, probably inevitably, in such a situation. Teaching thematically means that we generally only teach what is needed for that theme. This may only be *part* of a chemistry concept. Of course, a good curriculum returns to the concept later on and "completes" it. This means that the curriculum is inherently *spiral*. Each time a part of the concept is introduced, students learn it from a social or personal perspective. On the other hand, spiral, theme-based instruction can be a problem if instructors feel it is important to teach "all" of a given content area at once. Documenting the way this "spiral curriculum" works with wide-ranging content standards requires curriculum developers to trace the growth of concepts across units.

Table 2 shows how we have done this in our program by listing how *ChemCom* teaches periodicity over time. Different aspects of the concept are introduced only when that aspect is needed, but ultimately, the concept is covered completely.

Table 2. Chemistry in the Community and the Spiral Treatment of Periodic Properties

Unit Section	ChemCom Social or Personal Perspective	Concepts of Periodicity
2.A	The periodic table enables us to categorize element properties, helping us document the usefulness of certain elements.	Periodicity of properties, metallic vs. nonmetals, nuclear charge as the basis of atomic number and integrity of the table of elements
2.B	To obtain metals in a useful form from ore requires chemical reactions, including oxidation-reduction reactions.	Periodic trends in reactivity, oxidation-reduction behavior of the elements, protons and electrons equal in neutral atoms, ions formed by electron transfer
3.A	Different parts of petroleum are made of different alkanes that, despite their variety, are based on the same kinds of bonds.	Electron shells, valence electrons, and covalent bonding (Lewis structures)
6.A	The effects of ionizing radiation on human health, including disease and diagnostics, are based in the reactions of atomic nuclei.	Structure of the atom, isotopes, transmutation (changing) of elements through nuclear processes

Addressing Other Standards Through a "Relevant" Text

As suggested in Fig. 1, a text that covers personal and social perspectives also provides opportunities to address other standards, in particular the inquiry standard (Standard A), life science standards (Standard C), and earth and space science standards (Standard D). Chapters 4, 5, and 6 in this book deal more directly with these standards. Here, we note how our work to support chemistry instruction through *ChemCom* has allowed us to address aspects of those other standards. This is important within standards-based curricula, such as in Illinois, where state standards do require students to be tested on content outside of biology, chemistry, and physics.

The way in which thematic texts address nonchemistry standards is by the selection of the themes themselves. This means that in most cases, thematic teaching almost automatically touches on a broader set of standards (Brunkhorst 1997). For example, in Unit 7, students learn about the flow of energy from the Sun to Earth. This unit describes sunlight as a source of food energy in earth's ecosystems (NSES Standard C and ILS 12C), simultaneously providing a perspective for learning chemistry and also reinforcing the biology that the students learned the year before.

Of course, an inquiry text like *ChemCom* also addresses Standard A, learning about scientific inquiry and about learning through inquiry methods. In our case, the students in our program have already experienced work with an effective inquiry-based text, *BSCS: A Human Approach*. They learn to work from the "5E" inquiry model (engage, explore, explain, evaluate, and elaborate). To maintain this method of inquiry, we mapped the way that *ChemCom* can also be seen as using these same "5E" activities.

The "5E" mapping provides pedagogical continuity with the first year of our curriculum. But the thematic method of *ChemCom* also introduces a further kind of inquiry, by emphasizing that learning occurs purposefully through weeks of exploring content relevant to the solution of problems within society. Such design inquiry contrasts to disciplinary inquiry that is found in the development of science content understandings (Rudolph, 2005). Inquiry that shows how chemistry is useful also addresses Standard F directly.

A specific example of this comes from mapping the chemistry content of *ChemCom*'s Unit 1 with a student project to address the cause of a fish kill in a community (Table 3).

Table 3. Excerpts of Mapping of Design Inquiry Instruction With Content Knowledge, From Unit 1 of *ChemCom*

Chapter Section	Water Quality Perspective	Science Content
A.2 Uses of Water	Water use and amount	Water as a chemical compound and its use throughout daily life
B.4 Particulate View of Water	Water purity	Water in chemical mixtures and as a solvent
B.14 What Are the Possibilities?	Dealing with data	Using observations to form scientific predictions
D.1 Natural Water Purification	Water purification	Hydrologic cycle of water

The final aspect of inquiry learning that needs to be developed is in the development of student understanding of inquiry itself (Abraham, 2005). In this case, we chose to use a metacognitive strategy, where we gave assignments that require students to reflect on the content and the social or personal meaning of what they have experienced, especially in the laboratory. Our method was to take all three texts and embed their lab work in the Science Writing Heuristic (SWH) (Greenbowe and Hand, 2005; Hand and Keys, 1999; Keys et al., 1999). The detailed experimental procedures of many *ChemCom* activities were retained to guide students in their investigations; however, we also presented templates to the students that put the lab in the context of the SWH and also included specific questions that require students to sum up their understandings in the end. In many cases, these questions also reinforce the social and personal perspective that is behind the activity.

An example of the integration of all these inquiry strands and Standard F in a single *ChemCom* activity is found in Unit 1, "B.11: Water Testing" investigation. Here, students test water samples for the presence of dissolved ions. First, we have documented how well this matches with the idea of an *Explore* activity in the 5E model, alerting teachers and students to the presence of experiences that students will use as the basis of their learning. Second, our SWH template asks students to document beginning ideas by reflecting before the lab on what they have seen on TV, read about, or experienced firsthand about medical tests. After students perform the experiment and write claims that they can make based on evidence they have collected, a concluding question asks them to think again about the medical testing discussed earlier, now suggesting to the students how what they have learned may change how they think about medical tests. Third, this activity has been flagged for them as crucial to the content they will need to know in order to perform well on the benchmark assessment associated with each of the unit themes known as *Putting It All Together,* or PIAT, allowing for the personal perspective of the unit to be a foundation for their learning about ion testing.

Assessing Student Progress

Content testing in Illinois includes not only content, but also understanding both science inquiry and how science links to social and personal perspectives. Therefore, our assessments also must measure to what extent students have linked their knowledge to issues outside the classroom. Therefore, they include questions linked directly to the thematic organization of the units.

Our IDS provides a standards-based assessment program that is designed to support our approach to a vertically integrated sequence of courses. Students are assessed approximately every eight weeks throughout the academic year with both formative and summative exams. These exams are articulated with the Illinois Learning Standards and resemble, in both format and content, what students will encounter on the content-focused science exam taken on day two of the PSAE. The formative assessments provide data to analyze student progress toward meeting standards.

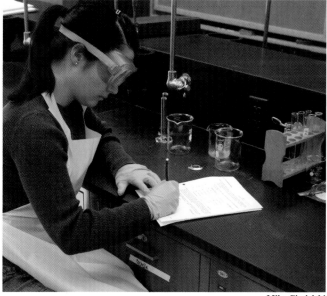

Mike Ciesielski

However, formative and summative assessments alone do not measure all we wish to know about student learning. We have also developed embedded benchmark assessments that are articulated with our vertically aligned scientific concepts and skills (see Table 1). Because all of the courses are built around the same key concepts and skills, teachers will be able to measure students' mastery of these skills both in individual courses, and over time, as the sequence of courses unfolds. Because the benchmark assessments are drawn from in-class work, they also capture how well students can *use* their emerging understandings. Each benchmark assessment identifies for the teacher the target concept or skill for a particular lesson, the specific student task that will be used to assess this skill, and the rubric to be used for grading the identified task. A common set of rubrics linked to our IDS concepts and skills provides consistent and clear student evaluation. In addition to the grading rubric, teachers are provided with examples of student work that correspond to meaningful differences in mastery of the concept or skill being assessed. In this way, assessment is built into the fabric of the entire program and is one of the essential supporting elements of classroom instruction.

The benchmark assessments also attend to student progress in areas related to Standard F. For example, in Unit 1 PIAT, we have identified the Nature of Science/Science as a Human Endeavor as the key concept for assessment.

The material in this lesson focuses on accumulating evidence to address a concern for society: clean water. This, therefore, emphasizes the nature of science as a data-gathering enterprise in service of human needs. The role of argument in bringing specific scientific facts to the attention of the community is also treated.

The teachers are then asked to look, in particular, at the way students use evidence about chemistry *and* the relevance of chemical concepts to the social question of the cause of a fish kill. In order to get full marks on the assessment, we suggest that students show that the data have meaning to society only if they are accepted as part of the argument. This underscores the need for students to develop an understanding of how science presents results that are *then* embraced (or not) by the community.

Other benchmark assessments are more conventional, ranging from proportional reasoning to questions of particulate representations of molecules. So, too, most of the formative and summative assessment questions are recognized as testing students on "standard" content. However, the presence of significant numbers of assessment items that link student learning to the social and personal perspectives of Standard F means that the standards are articulated to the tests.

Addressing the Needs of Students in Urban Districts with *ChemCom*

As suggested in the introduction, the first four issues that we find associated with teaching from a theme-focused text like *ChemCom* are likely to arise in almost all teaching contexts and with other thematic texts. But, in practice, teachers and curriculum developers have to address the link of a *particular* text to *particular* classrooms and students. Therefore, we wish to highlight an additional set of issues that are important in our program:

- students' low motivation to learn and lack of readiness for work in high school level science,
- potentially large percentages of English language learners and students with special learning needs,
- students with low reading levels, and
- creation of appropriate challenge levels for students who are highly capable but, because of school resources or curricula, do not have access to honors or advanced placement-level courses

The particular needs of students with low motivation and lack of prior knowledge are addressed in the use of relevant topics and activities as presented in *ChemCom*'s PIAT assignments. A high number of students enrolled in the urban high schools within this initiative come with minimal or *no* experience in science. Therefore, to draw on prior or current ideas about science, teachers must first *create* the experiences and have students generate ideas about their experiences. That is why activities that engage and explore (in the 5E model) are so important. If effective, these initial experiences engage students and assist in creating a need for learning the chemistry concepts. As a result, students do develop enough experiences to provide a basis for their learning. Equally important in engagement is the question of "relevance." Users of *ChemCom* and other thematic curricula have found that even when a scenario is not directly relevant to students, the chance to role play, problem solve, and argue about the social and personal meaning of a topic provides a powerful motivation for students, who often wind up embellishing their work to make it even more interesting.

Several strategies that are useful for all learners, but particularly for ELL and students with special needs, are described by Kimbrough and Cooper in chapter 13 of this book. We have also developed a consistent set based on our overall program and on the suggestions within *ChemCom*. The organization and use of *ChemCom* within this initiative highlights the following strategies to assist teachers in addressing students' needs:

Relevant activities and topics to engage students who are part of an urban environment (Examples include water quality and usage (Unit 1), fuels and transportation (Unit 3), and air quality (Unit 4));

- Collaborative group work (Students are involved in group work in many aspects of the curriculum, not just for lab activities. PIATs provide the best examples for such work, e.g., when students create a town council meeting with debates and presentations (Unit 1) and when they act as reporters for a school bus-idling policy (Unit 4));
- Conceptual connections within the material (*ChemCom*'s modeling matter activities require critiquing and creating visual representations of chemical activity making abstract concepts easier to understand);
- Spiral curriculum (The basis of *ChemCom* described earlier in the chapter, in which concepts are partially addressed within the context of a particular unit and then revisited either in greater depth or in application to a new context. Many concepts like atom economy and periodic relationships are woven throughout units);
- Varied forms of assessment (Incorporation of literacy strategies and the SWH instruction help students manage their learning, reading, and writing. Even for students who cannot write well, the SWH guides their thinking processes (Hand, 2006; Burke et al 2006), while they use pictorial diagrams to show experimental designs); and
- Responsibility to a community (Several activities emphasize the importance of understanding chemical concepts when making decisions in communities. The understanding of chemistry is not just for scientists but is often needed by business owners, government agencies, and neighborhood action groups. This is also exemplified in the town council meeting activities at the end of Unit 1).

Students who struggle to read and write in English are supported with collaborative group work and the SWH and are able to express their ideas in various ways as they build their literacy. As Klentschy and Molina-De La Torre (2003) have observed, when instruction successfully improves both literacy and understanding of science concepts, "students have personal, practical motivation to master language as a tool that can help them answer their questions about the world around them."

Finally, students who are highly capable can be assisted in their own further development. When placed in groups with other students of varying ability, highly capable students can be coached in becoming experts in particular topics by conducting more in-depth reporting or research. *ChemCom*'s ChemQuandries can provide guidance for such supplemental work. For example, in ChemQuandry 3 found in Unit 2 students are told that a U.S. nickel is composed of an alloy of nickel and copper—specifically 25% Ni and 75% Cu. They are asked to consider the appearance of a nickel and how this given composition might be surprising. Students are then asked to determine the difference between an alloy (a "solid solution" of copper and nickel atoms) and a simple mixture of powdered copper and powdered nickel. A highly capable student might be then asked to research the compositions of various coins (domestic and foreign) and report findings to the class.

Conclusion

We have discussed the challenges and opportunities that come to curriculum developers and teachers using materials that embed learning in social and personal perspectives—essentially, materials that implement Standard "F" as a key part of their planning. We have described this in the context of one particular effort at providing a diverse, urban high school district with a standards-based curriculum using *ChemCom*. Similar opportunities (and challenges) are present in other texts such as *Active Chemistry* and *Living by Chemistry*, which both include unit themes that organize and encapsulate the learning of chemistry.

References

Abraham, M. R. Inquiry and the Learning Cycle Approach. In *Chemists' Guide to Effective Teaching*; Pienta, N. P., Cooper, M. M., Greenbowe, T. J., Eds.; Prentice Hall: Upper Saddle River, NJ, 2005; pp 41–52.

American Chemical Society. *Chemistry in the Community,* 5th ed.; W. H. Freeman: New York, 2006.

Biological Sciences Curriculum Studies. *Biology: A Human Approach*, 3rd ed.; Kendall Hunt: Dubuque, IA, 2006.

Bridgeland, J. M.; DiIulio, J. J. Jr.; Morison, K. B., *The Silent Epidemic: Perspectives of High School Dropouts;* Civic Enterprises: Washington, DC, 2006. http://www.gatesfoundation. org/nr/downloads/ed/TheSilentEpidemic3-06FINAL.pdf (accessed March 2008).

Brunkhorst, B. "Grounding" Chemistry with Earth and Space Science. In *Chemistry in the National Science Education Standards;* American Chemical Society: Washington, DC, 1997; pp 53–61.

Burke, K. A.; Greenbowe, T. J.; Hand, B. M. Implementing the science writing heuristic in the chemistry laboratory. *J. Chem. Educ.* 2006, *83*, 1032–1038.

Eisenkraft, A. *Active Physics;* It's About Time: Armonk, NY, 2004.

Freebury, G.; Eisenkraft, A. *Active Chemistry;* It's About Time: Armonk, NY, 2006.

Greenbowe, T. J.; Hand B. M. Introduction to the Science Writing Heuristic. In *Chemists' Guide to Effective Teaching*; Pienta, N. P.; Cooper, M. M.; Greenbowe, T. J., Eds.; Prentice-Hall: Upper Saddle River, NJ, 2005; pp 140–154.

Hand, B. Using the Science Writing Heuristic to Promote Understanding of Science Conceptual Knowledge in Middle School. In *Linking Science & Literacy in the K–8 Classroom*; Douglas, R.; Klentschy, M. P.; Worth, K., Eds.; NSTA Press: Arlington, VA, 2006; pp 117–125.

Hand, B.; Keys, C. W. Inquiry Investigation: A New Approach to Laboratory Reports. *Sci. Teach.* 1999, *66*, 27–29.

Illinois State Board of Education. *Illinois Learning Standards for Science.* Springfield, IL: Illinois State Board of Education, 1997, http://www.isbe.state.il.us/ils/science/standards. htm (accessed March 2008).

Keys, C. W.; Hand, B.; Prain, V.; Collins, S. Using the Science Writing Heuristic as a Tool for Learning from Laboratory Investigations in Secondary Science. *J. Res. Sci. Teach.* 1999, *36*, 1065–1084.

Klentschy, M.; Molina-De La Torre, E. Students' science notebooks and the inquiry process. In *Border Crossing: Essays on Literacy and Science;* Saul, W., Ed. NSTA Press: Arlington, VA, 2003; pp 1–23.

National Research Council. *National Science Education Standards.* National Academies Press: Washington, DC, 1996.

Rudolph, J. L. Inquiry, Instrumentalism, and the Public Understanding of Science.Sci. Educ.2005,803–821.

Stacy, A. *Living by Chemistry*, Key Curriculum Press: Emeryville, CA, 2006.

Chapter 9

Content Standards for the History and Nature of Science

by Seth C. Rasmussen, Carmen Giunta, and Misty R. Tomchuk

Seth C. Rasmussen *is an associate professor of chemistry at North Dakota State University in Fargo, ND. In addition to ongoing research in materials chemistry, he maintains an active interest in the history of chemistry and has organized two national symposia on its use in chemical education. Contact e-mail: seth.rasmussen@ndsu.edu*

Carmen Giunta *is a professor of chemistry at Le Moyne College in Syracuse, NY. A physical chemist by training, he is particularly interested in applying the history of chemistry to chemical education. He maintains the Classic Chemistry Web site: http://web.lemoyne. edu/~giunta/. Contact e-mail: giunta@lemoyne.edu*

Misty R. Tomchuk *teaches physical science at Fargo North High School in Fargo, ND. She spent 5 years doing chemistry research for North Dakota State University and currently teaches science to an at-risk population. Contact e-mail: tomchum@fargo.k12.nd.us*

Introduction

The importance of history in chemical education. Before the 1950s, the inclusion of the history of science in the teaching of science was commonplace. Unfortunately, this is no longer the case. Possible explanations for this omission include the views that science history is no longer a legitimate subject in science education, that science history does not contribute to learning the technical aspects of science (Brush, 1974), or that it is difficult to fit into the topic loads of current science classes (Matthews, 1994; Rasmussen, 2007). Over the last several decades, however, the history of science has enjoyed a resurgence, along with a growing awareness regarding the important roles it plays in science education:

- History promotes better comprehension of scientific concepts and methods (Matthews, 1994; Schwartz, 1977).
- History illustrates the importance of individual thought and creativity in the development of science (de Carvalho and Vannucchi, 2000; Kauffman, 1987; Matthews, 1994; Schwartz, 1977).
- History of science is intrinsically worthwhile (Kauffman, 1987; Matthews, 1994).
- History is necessary to understand the nature of science (de Carvalho and Vannucchi, 2000; Giunta, 1998; Herron et al., 1977; Kamsar 1987; Kauffman, 1987, 1989; Matthews, 1994).
- History counteracts the dogmatic view of science commonly found in texts and classes (de Carvalho and Vannucchi, 2000; Giunta, 2001; Herron et al., 1977; Kauffman, 1987, 1989; Matthews, 1994; Schwartz, 1977).

- History humanizes the subject matter of science, making it less abstract and more engaging for students (Herron et al., 1977; Kamsar, 1987; Kauffman, 1987, 1989; Matthews, 1994; Schwartz, 1977).
- History shows the connections among chemical disciplines (Kauffman, 1987; Matthews, 1994).
- History allows one to more easily identify pseudoscience (Rasmussen, 2007).

In fact, it can be argued that the history and philosophy of science are an integral part of scientific knowledge and, therefore, must be included in science education (de Carvalho and Vannucchi, 2000). As such, the inclusion of historical and philosophical aspects of science in high school courses has been a key recommendation of science-teaching research studies (de Carvalho and Vannucchi, 2000; Matthews, 1994). This is echoed in content standard G of the *National Science Education Standards* (NSES): "all students should develop [an] understanding of science as a human endeavor, [the] nature of scientific knowledge, and historical perspectives" (National Research Council, 1996).

The importance of chemistry in the history and nature of science. Chemistry is often referred to as "the central science" because it overlaps and bridges the other sciences, including those of physics and biology. As such, the inclusion of chemistry in the history of science is especially important. And while the history of science, in general, is important in science education, all too often, these historical inclusions tend to focus on the history of physics or biology, rather than on the history of chemistry. It is thought by some that this is at least partially due to the fact that in comparison to other scientists, chemists display relatively little interest in the history of their own subject, and the general state of research in the history of chemistry is not as strong as that of the history of physics (Kauffman, 1989; Matthews, 1994). Contrary to these views, however, is the fact that numerous authors have championed the inclusion of history of chemistry into chemistry education (Kauffman, 1989). Not only does it give students a greater comprehension and appreciation of chemistry itself, but it also illustrates better than any other scientific discipline how all science, medicine, and technology are intertwined and related (see discussion in *Chemistry, history, and interdisciplinarity* below).

The Proper Use of History in Chemical Education

The dangers of beautifying or simplifying history. Introducing historical materials is often done selectively, as the primary purpose is to use these materials to teach modern theories and techniques more effectively. However, overly selective use of history may result in a series of fascinating anecdotes rather than factual history (Brush, 1974). Students may gain no real understanding of the problems that concerned past scientists, the context in which they worked, or how they convinced their contemporaries to accept new ideas (Matthews, 1994).

One dangerous trend along these lines is to start with historical fact, but to then remove all information deemed unnecessary, too confusing, or contradictory to the idea being taught. The problem, however, is that this simplified product now gives students a skewed (if not completely incorrect) view of history. This treatment of history can make it seem that the original discovery or theory was without flaws and that the modern form is identical to the original, or at least in perfect agreement with accepted thought. Although this simplified version is easier for the student to incorporate into the modern chemistry being learned, it removes the important fact that our modern theories and concepts did not develop in a linear, logical style. Simplified versions of history run the risk of furthering the student's misconception that modern chemistry is a finished and absolute product unchangingly etched in stone (Giunta, 2001; Kauffman, 1987, 1989). Such simplified history is commonly found in contemporary textbooks and contributes to a distorted form of scientific history that is more myth and legend than true history (Brush, 1974). If we teach science as the pursuit of truth and fact, should this not also hold true of our treatment of our past heritage as well?

True history vs. idealized history. While idealized history may make a more digestible story, it not only gives a false view of history, but also removes a number of opportunities to use history as a powerful teaching tool. A true historical approach that includes all the error, approximation, and human foibles allows students to witness the reality of science at work (Schwartz, 1977). Here, students can see that while intellect and education are important, so too are enthusiasm, optimism, an appetite for hard work, and a bit of luck. Likewise, a true historical approach recognizes imagination and gives students better recognition of their own creative abilities as they learn that intuition, as well as logic, is a legitimate approach to problem solving (Kauffman, 1987; Schwartz, 1977).

One of the concerns expressed about exposing students to true history, rather than an idealized, sanitized version, is that it may turn impressionable students away from chemistry by letting them see that chemists do not always behave as rational, open-minded investigators who proceed logically, methodically, and unselfishly toward the truth (Kauffman, 1989). In particular, by taking an open, honest look at some of the most revered figures in chemistry, we may somehow tarnish their reputations and reduce students' admiration for these scientists and their accomplishments (Schwartz, 1977). However, one could argue that this is just as valid a reason to include the full, honest truth in history. As educators, it is becoming more and more common to witness students begin their study of science with the attitude that there is no way that they can master the subject. Such students feel that such accomplishments are far too hard for a "normal" student such as themselves and that to succeed in science requires exceptional intellectual abilities. Recognizing that great figures of chemistry were human beings with strengths and weaknesses not all that different from themselves can give students the confidence to try. For such students, true history can illustrate the number of times great discoveries have been made by those with average abilities, poor training, or faulty logic and can, just as importantly, show that such discoveries are rarely made by one scientist alone, but that such accomplishments were also dependent on the work, theories, and insight of other contributing scientists.

JupiterImages

Science as a Human Endeavor

Humanizing science. The primary goal of NSES content standard G is that all students should develop an understanding of science as a human endeavor (National Research Council, 1996). This is important to counteract the sterile, cold image often associated with science. Students are typically more receptive to the subject when they can visualize people in science. The application of the historical approach, with its emphasis on people and society, can be an excellent tool to place chemistry in perspective as a human activity (Herron et al., 1977). As stated by Jaffe (1955), history can show that "Inextricably tied to these world-shaking advances was an even greater story—the human one—the saga of men [*sic*] groping for causes and struggling to frame laws; of men [*sic*] leading intellectual revolutions and fighting decisive battles in laboratories. Here was meaning, light, inspiration, life."

Illustrating the diversity of scientists and the international character of science is crucial to emphasize in teaching about science as a human endeavor (Jaffe, 1955; Kauffman, 1989). Teachers must be careful to use historical materials that undermine, rather than reinforce, the tendency of many students to view science as a product of men from the United States and Europe. The very words from Jaffe (1955) quoted above limit that enterprise to men. Many works of history of chemistry do, in fact, emphasize the achievements of men of European descent, because social factors limited the participation of women and of many non-European ethnic groups in science. Intellectual honesty requires acknowledgment of this historical reality, whose effects still influence the present. In light of this reality, selecting examples of women and other underrepresented groups who made significant contributions to chemistry despite disadvantages, is vitally important. The Eurocentric male view can, and should, be dispelled as students come to understand that no gender, country, or culture has a monopoly on discovery (Kauffman, 1989) and that many of chemistry's greatest discoveries originated in the Middle East, Egypt, and Asia.

The biographical approach. One of the easiest methods to convey science as a human endeavor is through the biographical approach, which can also be one of the most inspiring to students (Kauffman, 1989). The benefits of a biographical approach in teaching chemistry have been widely recognized (Kauffman, 1971), and there are a number of possible methods for incorporating scientific biography into chemistry classes. Short biographical statements are often included in textbooks, and, while they are typically examples of idealized history as discussed above, they offer a simple starting point to introduce a particular scientist. The life and work of the scientist can then be viewed in more depth through classical lecture or teacher-led discussion (Jaffe, 1955) or can lead to student-oriented projects, including short dictionary/encyclopedia-style writing assignments (Jensen, 2001), term papers (Kauffman, 1971), class presentations, or even cooperative group research projects and presentations (Carroll and Seeman, 2001). The biographical approach also has the benefit of availability of an abundance of quality resources, including standard texts (Bowden, 1997; Jaffe, 1976), online databases and biographies (Burda, 2008; Burke, 2008; Mainz, 2008; Nobel Foundation, 2008; Chemical Heritage Foundation, 2008) and video sources (Djerassi and Hoffmann, 2003; Smith et al., 2007).

History and the Nature of Scientific Knowledge

Chemistry, history, and interdisciplinarity. Whereas science as a human endeavor appears as a content standard throughout K–12 education, standards on the nature of science do not appear until the middle grades. Fortunately, this is still before students begin to take disciplinary science courses. Interdisciplinarity is an increasingly important aspect of the current practice of science, so it is desirable that students start to encounter material on the nature of science before the idea of distinct scientific subjects becomes ingrained. As students learn about the nature of scientific inquiry (observation, experimentation, and the like) and the vast range of phenomena to which these methods of inquiry can be applied (plants, planets, pendula, and many, many more), the world of science appears without disciplinary boundaries. Eventually, students learn about scientific disciplines—and for good reason. Specialization is inevitable in science education, and those students who continue to study science in college and beyond will be exposed to ever more of it. Yet even as students begin to be exposed to distinct scientific specialties, it is worth pointing out that nature respects no disciplinary boundaries and that much contemporary science is done by interdisciplinary teams.

Much of the instruction on the nature of science can and ought to be done through hands-on activities (see chapter 4 on inquiry learning and chapter 7 on laboratory instruction). Historical material can also be profitably employed. Biographical materials that illustrate science as a human endeavor can also represent science as an interdisciplinary endeavor. Because chemistry (the "central science") has long shared borders with several scientific disciplines, there are many examples of scientists who contributed to chemistry as well as other scientific fields.

The individuals who made important contributions to chemistry include numerous examples of physicians, geologists, physicists, and those scientific generalists of yesteryear, natural philosophers. Antoine Lavoisier contributed significantly to physiology as well as chemistry. Henry Cavendish both discovered hydrogen and determined the mass of Earth. John Dalton formulated an atomic theory of matter and described color blindness (from personal experience). Ernest Rutherford, who disparaged sciences other than physics, received a Nobel Prize in chemistry. These historical examples illustrate that the permeability of scientific boundaries is not a modern phenomenon. They provide historical precedent for the multidisciplinary fields in which today's chemists participate—areas such as energy, materials, biotechnology, climate change, and nanotechnology.

Case studies in how science works. Historical material can shed light on other aspects of the nature of science besides interdisciplinarity. It is worth emphasizing to students that the importance of empirical testing in science does not imply a single "scientific method," let alone a method that precludes imagination, creativity, passion, or luck (see chapter 3 on unifying themes, particularly its treatment of evidence, models, and explanation.). Even though observation and experiment are what ultimately lead to acceptance by the scientific community, there are almost as many sources of scientific ideas, explanations, analogies, and questions as there are scientists. Derry's *What Science Is and How It Works* (1999) is a good primer for science teachers on the nature of science, and it treats several historical cases, including some in chemistry. The role of luck in science is emphasized in Roberts' *Serendipity: Accidental Discoveries in Science* (1989), a worthwhile book for both instructors and older students.

Scientists' descriptions of their own discoveries can also shed light on the nature of science. First-hand descriptions must be carefully selected (and often annotated or supplemented) to be understandable by students and useful in the classroom. Such accounts can reveal the relationship between what an investigator observed and what he or she concluded, and they can also reveal lines of inquiry (fruitful and otherwise) prompted by a preliminary observation, and even instances of speculation that proved to be mistaken or fortuitous. Some of the annotated papers in the "Elements and Atoms" portion of the Classic Chemistry Web site (Giunta, 2008) are suitable for high school students. The descriptions of Lavoisier's work on the nature of water and Becquerel's experiments on penetrating rays from uranium salts (i.e., radioactivity) are sufficiently clear, rich, and brief to warrant close reading. Tracing the evolution of some chemical concepts can shed light on current understanding of those concepts. Atomic structure is certainly a topic amenable to a "history of ideas" treatment: seeing successive models incorporate new knowledge as it became available both illustrates how scientific knowledge can be constructed and highlights the evidence behind our current understanding.

Historical Perspectives

Chemistry and society. Historical perspectives on science can illustrate the pervasiveness and variety of interactions of science and technology with social, political, and economic history. Issues of science and society necessarily encompass interdisciplinary approaches writ large. That is, we are no longer talking about interdisciplinarity simply within natural science but between natural science, social science, and the humanities. We hope that teachers of all of these subjects will point out relevant connections to the other subjects. Because our readers are chemistry teachers, we limit ourselves to connections they can make to other subjects; at the same time, we encourage chemistry teachers to work with their colleagues in other subjects to mutually reinforce these connections.

Chemists are, understandably, concerned with the public image of their profession. The persistent association of chemistry with pollutants and other hazardous materials presents a stark contrast to the vision statement of the American Chemical Society (Raber, 2006), "Improving people's lives through the transforming power of chemistry." Chemistry teachers can show chemistry as a relevant and beneficial enterprise by providing examples of how it affects society past and present (see chapter 8). However, in providing a beneficent view of chemistry, the temptation to gloss over real problems must be avoided (see *True history vs. idealized history* above). Balanced accounts are both more interesting and more intellectually honest than simplistic accounts that portray chemists as heroes or villains.

One particularly rich episode from 20th-century chemistry involves chlorofluorocarbons (CFCs) and the ozone layer. Chemists play pivotal roles throughout this story, from the invention of CFCs as nontoxic and nonflammable refrigerants, to detecting alarming quantities of them still in the atmosphere years after their release, to unraveling the complex interactions that cause CFCs eventually to catalyze destruction of ozone in the stratosphere. An additional feature that makes this story a particularly relevant case of science and society is the ultimate action of governments around the world to regulate the hazard. In short, chemists made the problem, chemists identified the problem, and chemists provided governments with information that permitted solution of the problem.

Interactions between chemistry and society need not have such global significance to capture students' attention, though. Multiple connections can be made between chemistry and the ostensibly more "creative" field of art. Chemistry, of course, is responsible for the materials artists use today (and in some cases, have used for centuries) to give shape and color to their visions. Pigments, sculpting media, and conservation and restoration of art objects are topics that lie at the intersection of chemistry, art, and history. Kafetzopoulos et al. (2006) describe hands-on activities that link chemistry and art. Kelly et al. (2001) describe an interdisciplinary college course that bridges art and science; their article contains references that can be useful in the high school classroom as well.

JupiterImages

Teaching history and teaching through history.

Although history of science is itself a prominent part of one of the National Science Education Standards, science teachers strive to impart information and interest in science. Thus, many consider historical material to be supplemental enrichment at best and a distraction at worst. There are, however, several ways to include historical material in the service of learning objectives more directly related to chemistry content and problem solving. One such way was hinted at in the previous paragraph: material that stimulates students' interest in chemistry can motivate the effort and attention required to master chemistry content. Art may attract some students while glamour, sex, or crime may attract others. Although some of these subjects ought to be broached carefully, there are excellent popular books on chemical aspects of such subjects that have the capacity of capturing the interest of some adolescents—and adults as well (Emsley, 2004, 2005).

Stimulating interest serves chemistry learning objectives indirectly. Historical material can also be used in direct support of chemistry content and problem-solving skills by basing quantitative problems on historical data. When teaching Boyle's law, for example, Boyle's data can be displayed, plotted, and fit. When teaching how to determine empirical formulas on the basis of chemical analysis, data from one of the classic 19th-century analytical chemists can be employed. A set of quantitative exercises based on historical data developed mainly for introductory college chemistry can be found as "Classic calculations" at the Classic Chemistry Web site (Giunta, 2008), and some of these can be used or adapted for use in high school.

Conclusions

We hope that the above discussion and arguments bring to light both the importance of the inclusion of the history of chemistry in chemistry education and many of the ways that this can be accomplished. The incorporation of historical anecdotes in a standard chemistry course can make the course more interesting and relevant—especially for nonscience students (Herron et al., 1977). In addition, the time taken in the classroom for the addition of chemical history provides a "thinking floor" for future chemists to find their place in society and, most importantly, creates interest and provokes thought about the scientific endeavor for the majority of pupils who will not become scientists (Kamsar, 1987). If we want our students to have a real conceptual understanding of scientific progress and practice, then we must go beyond the regurgitation of experimental details (Niaz and Rodriguez, 2001); the use of history allows us to accomplish this goal while providing a richer learning environment for our students.

Recommended Readings

Derry, G. N. *What Science Is and How It Works*; Princeton University Press: Princeton, NJ, 1999. Addresses many facets of the nature of science, including several historical cases, some in chemistry. Aimed at college nonscience majors, so suitable for teachers and education students.

Emsley, J. *Vanity, Vitality, and Virility: The Science Behind the Products You Love to Buy*; Oxford University Press: New York, 2004. Chemical aspects of selected cosmetics, personal care products, dietary supplements, and drugs described for a general audience.

Emsley, J. *The Elements of Murder*; Oxford University Press: New York, 2005. Toxic chemical elements and their uses and abuses over the years, described for a general audience.

Jaffe, B. *Crucibles: The Story of Chemistry, From Ancient Alchemy to Nuclear Fission*, 4th ed., Dover: New York, 1976 (first edition 1930). A set of biographies of famous chemists. In its day, it was popular and award-winning for humanizing chemists; the absence of women and other underrepresented groups in its pages is a shortcoming.

Kafetzopoulos, C.; Spyrellis, N.; Lymperopoulou-Karaliota, A. The Chemistry of Art and the Art of Chemistry. *J. Chem. Educ.* 2006, *83*, 1484–1488. Describes hands-on classroom activities that connect chemistry and art.

Kelly, C.; Jordan, A.; Roberts, C. Finding the science in art. *J. Coll. Sci. Teach.* 2001, *31*, 162–166. Describes an interdisciplinary college course that bridges art and science; its references can be useful to the high school classroom as well.

Matthews, M. R. *Science Teaching, The Role of History and Philosophy of Science*; Philosophy of Education Research Library; Routledge: New York, 1994. Excellent review of using the history and philosophy of science in science education.

Ringnes, V. Origin of the names of chemical elements. *J. Chem. Educ.* 1989, *66*, 731–736. Source of interesting historical information on the elements.

Roberts, R. M. *Serendipity: Accidental Discoveries in Science*; Wiley: New York, 1989. Source of anecdotes that illustrate how luck can be an important ingredient in discoveries and inventions.

Recommended Web Sites and Other Media

The intersection of history of chemistry and chemical education is an area to which high school teachers can contribute, and the Internet makes it easier to distribute that work. One of the first Internet sites to republish classic historical papers, and still one of the richest, was created by a chemistry teacher, John Park. A more recent Web site devoted to discoveries and names of the chemical elements was prepared by another high school teacher, Dave Trapp, based on an article by Vivi Ringnes (1989).

Bowden, M. E. *Chemical Achievers: The Human Face of Chemical Sciences*; Chemical Heritage Foundation: Philadelphia, 1997; Web version can be found at http://www.chemheritage.org/classroom/chemach/index.html (accessed March 18, 2008). Pictures and short biographical sketches of prominent chemists. Underrepresented groups are included, but not as a primary emphasis.

Burda, G. A., Ed., Women in Chemistry: Her Lab in Your Life. http://www.chemheritage.org/women_chemistry/index.html (accessed March 18, 2008). Biographical sketches of women in chemistry, past and present.

Burke, B. A., Ed., JCE Online: Biographical Snapshots of Famous Women and Minority Chemists. http://jchemed.chem.wisc.edu/JCEWWW/Features/eChemists/index.php (accessed March 18, 2008). Biographical sketches of women in chemistry and chemists from ethnic minorities.

Chemical Heritage Foundation. Science Alive: Percy Julian. http://www.chemheritage.org/scialive/julian/index.html (accessed March 18, 2008). Classroom materials and exercises based on the life and work of a 20th century African American chemist.

Djerassi, C.; Hoffmann, R. Oxygen [DVD]; Wisconsin Initiative for Science Literacy, 2003. Recorded production of a play about the discovery of oxygen that explores questions of credit and motivation; written by two prominent chemists; a teacher's guide is also available.

Giunta, C. J. Classic Chemistry. http://web.lemoyne.edu/~giunta/ (accessed March 18, 2008). "Elements and Atoms" section contains papers annotated to illustrate process of discovery; "Classic Calculations" includes quantitative exercises based on historical data.

Mainz, V. Chemical Genealogy. http://www.scs.uiuc.edu/~mainzv/Web_Genealogy/ (accessed March 18, 2008). Database of chemical researchers' scientific "lineages," illustrating mentor-student connections and influences. Each entry contains a short biography with references to additional information.

Nobel Foundation. All Nobel Laureates in Chemistry. http://nobelprize.org/nobel_prizes/chemistry/laureates/ (accessed March 18, 2008). Biographies of Nobel laureates in chemistry and descriptions of their work, including descriptions in their own words and some classroom materials.

Park, J. ChemTeam: Classic Papers Menu. http://dbhs.wvusd.k12.ca.us/webdocs/Chem-History/Classic-Papers-Menu.html (accessed March 18, 2008). Primary sources of many classic chemical discoveries in the words of the original researchers.

Smith, L.; Lyons, S.; Quade, D.; Santiago-Hudson, R.; Vance, C. B. *Forgotten Genius* [DVD]; WGBH, 2007. Features the life and work of 20th-century African-American chemist Percy Lavon Julian.

Trapp, D. Discovery and Naming of Chemical Elements. http://homepage.mac.com/dtrapp/Elements/elements.html (accessed March 18, 2008). Source of interesting historical information on the elements.

References

Bowden, M. E. *Chemical Achievers: The Human Face of Chemical Sciences*; Chemical Heritage Foundation: Philadelphia, 1997; Web version at http://www.chemheritage.org/classroom/chemach/index.html (accessed March 18, 2008).

Brush, S. G. Should the History of Science Be Rated X? *Science* 1974, *183*, 1164–1172.

Burda, G. A., Ed., Women in Chemistry: Her Lab in Your Life. http://www.chemheritage.org/women_chemistry/index.html (accessed March 18, 2008).

Burke, B. A., Ed., JCE Online: Biographical Snapshots of Famous Women and Minority Chemists. http://jchemed.chem.wisc.edu/JCEWWW/Features/eChemists/index.php (accessed March 18, 2008).

Carroll, F. A.; Seeman, J. I. Placing Science Into Its Human Context: Using Scientific Autobiography to Teach Chemistry. *J. Chem. Educ.* 2001, *78*, 1618–1622.

Chemical Heritage Foundation. Science Alive: Percy Julian. http://www.chemheritage.org/scialive/julian/index.html (accessed March 18, 2008).

de Carvalho, A. M. P.; Vannucchi, A. I. History, Philosophy, and Science Teaching: Some Answers to "How?" *Science & Education* 2000, *9*, 427–448.

Derry, G. N. *What Science Is and How It Works*; Princeton University Press: Princeton, NJ, 1999.

Djerassi, C.; Hoffmann, R. *Oxygen* [DVD]; Wisconsin Initiative for Science Literacy, 2003.

Emsley, J. *Vanity, Vitality, and Virility: The Science Behind the Products You Love to Buy*; Oxford University Press: Oxford; New York, 2004.

Emsley, J. *The Elements of Murder*; Oxford University Press: New York, 2005.

Giunta, C. J. Classic Chemistry. http://web.lemoyne.edu/~giunta/ (accessed March 18, 2008).

Giunta, C. J. Using History to Teach Scientific Method: The Case of Argon. *J. Chem. Educ.* 1998, *75*, 1322–1325.

Giunta, C. J. Using History to Teach Scientific Method: The Role of Errors. *J. Chem. Educ.* 2001, *78*, 623–627.

Herron, J. D.; Boschmann, E.; Kessel, W.; Lokensgard, J.; MacInnes, D. The Place of History in the Teaching of Chemistry. *J. Chem. Educ.* 1977, *54*, 15–16.

Jaffe, B. Using the History of Chemistry in Our Teaching. *J. Chem. Educ.* 1955, *32*, 183–185.

Jaffe, B. *Crucibles: The Story of Chemistry, From Ancient Alchemy to Nuclear Fission*, 4th ed. Dover: New York, 1976.

Jensen, W. B. Aaron Ihde's Contributions to the History of Chemistry. *Bull. Hist. Chem.* 2001, *26*, 24–32.

Kafetzopoulos, C.; Spyrellis, N.; Lymperopoulou-Karaliota, A. The Chemistry of Art and the Art of Chemistry. *J. Chem. Educ.* 2006, *83*, 1484–1488.

Kamsar, J. W. Utilizing a Historical Perspective in the Teaching of Chemistry. *J. Chem. Educ.* 1987, *64*, 931–932.

Kauffman, G. B. Teaching the History of Science: A Biographical Approach. *J. Coll. Sci. Teach.* 1971, *1*, 26–28.

Kauffman, G. B. History of Chemistry *J. Chem. Educ.* 1987, *64*, 931–933.

Kauffman, G. B. History in the Chemistry Curriculum. *Interchange* 1989, *20*, 81–93.

Kelly, C.; Jordan, A.; Roberts, C. Finding the science in art. *J. Coll. Sci. Teach.* 2001, *31*, 162–166.

Mainz V. Chemical Genealogy. http://www.scs.uiuc.edu/~mainzv/Web_Genealogy/ (accessed March 18, 2008).

Matthews, M. R. *Science Teaching, The Role of History and Philosophy of Science*; Philosophy of Education Research Library; Routledge: New York, 1994; pp 48–82.

National Research Council. *National Science Education Standards*; National Academy Press: Washington, DC, 1996; Web version can be found at http://www.nap.edu/openbook.php?isbn=0309053269 (accessed March 18, 2008).

Niaz, M.; Rodriguez, M. A. Do We have To Introduce History and Philosophy of Science or Is It Already 'Inside' Chemistry? *Chem. Educ. Res. Pract.* 2001, *2*, 159–164.

Nobel Foundation. All Nobel Laureates in Chemistry. http://nobelprize.org/nobel_prizes/chemistry/laureates/ (accessed March 18, 2008).

Raber, L. R. ACS Launches New Vision. *Chem. Eng. News* 2006, *84*, 52–53.

Rasmussen, S. C. The History of Science as a Tool to Identify and Confront Pseudoscience. *J. Chem. Educ.* 2007, *84*, 949–951.

Ringnes, V. Origin of the Names of Chemical Elements. *J. Chem. Educ.* 1989, *66*, 731–736.

Roberts, R. M. *Serendipity: Accidental Discoveries in Science*; Wiley: New York, 1989.

Schwartz, A. T. The History of Chemistry: Education for Revolution. *J. Chem. Educ.* 1977, *54*, 467–468.

Smith, L.; Lyons, S.; Quade, D.; Santiago-Hudson, R.; Vance, C. B. *Forgotten Genius* [DVD]; WGBH, 2007.

Chapter 10

Professional Development of Chemistry Teachers

by Mickey Sarquis and Lynn Hogue

Mickey Sarquis, *professor of chemistry and biochemistry and director of the Center for Chemistry Education at Miami University, is an internationally recognized leader in chemistry and science education because of her pioneering chemistry-based teacher enhancement and curriculum development. She has received numerous teaching and service awards, including the Chemical Manufacturers Association National Catalyst Award for Excellence in Chemistry Teaching, and has published more than 60 books, monographs, chapters, and articles. Contact e-mail: sarquiam@muohio.edu*

Lynn Hogue, *associate director of Miami University's Center for Chemistry Education, has served as teacher programs director and lead instructor for more than 50 funded National Science Foundation and Ohio Board of Regents programs. A science educator for 30 years, she is a coauthor of teacher resource books, including* Investigating Solids, Liquids, and Gases with TOYS *and* Science Projects for Holidays Throughout the Year. *She has assisted school districts throughout the country with updating their science curriculum and presenting teacher inservices. Contact e-mail: hoguelm@muohio.edu*

Introduction

This chapter is written for the professional development provider and offers practical guidance, as well as supporting theory, to inform and improve professional development of chemistry teachers. Inservice teachers will benefit from reading this chapter and learning what best practices and aspects to look for when selecting a professional development opportunity. Throughout the chapter, "Less emphasis on …" and "More emphasis on …" examples highlight the nature of professional development envisioned in the *National Science Education Standards* (NRC, 1996).

The National Research Council (1996) and others (U.S. Department of Education, 2000; American Chemical Society, 2004; Loucks-Horsley et al., 1996) call for a professional development continuum that provides coherent and integrated opportunities for teachers to build and extend their content knowledge, pedagogical skills, and pedagogical content knowledge, while promoting professionalism and "understanding and ability for lifelong learning." (NRC, 1996) Many currently available professional development opportunities for chemistry teachers do not provide the broad spectrum of activities that are necessary for growth and called for in the *National Science Education Standards*. Additionally, many opportunities focus exclusively on content or pedagogy or attempt to teach content and pedagogy as two separate entities, rather than the richly intertwined elements that they are. Most professional development completely ignores the element of pedagogical content knowledge.

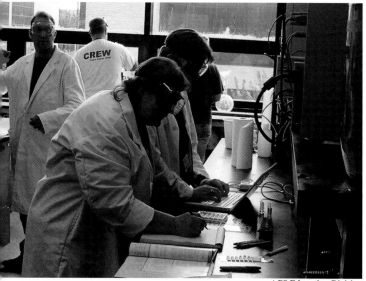
ACS Education Division

It is important to clarify that content knowledge means more than the acquisition of facts, vocabulary, laboratory skills, or mathematical algorithms. Content knowledge encompasses building mastery of the conceptual underpinnings of the subject matter that extend throughout the discipline. It means being able to identify the "big ideas" that drive science (see chapter 3) and to use these ideas to assist students in discovering the links between topics as we seek to provide opportunities for them to forge their own conceptual understanding. Finally, we should not forget that content knowledge also includes the dynamic of the nature of science as a process; educators must understand and be able to facilitate science processes such as questioning, observation, prediction, and argumentation in their classrooms (see chapter 4 on inquiry).

The building of pedagogical skills is equally important to the construction of content knowledge, particularly as a means of providing student-centered learning experiences, including inquiry, questioning to advance understanding, and science argumentation. Teachers need to have a thorough understanding of the process of scientific inquiry and develop ways to guide their students in this process. Dyasi and Dyasi (2004) state that "just as with student learning, continuing science education for teachers must center on firsthand experience with the ordinary physical world." The effective science teacher actively guides students to help them understand what they are doing and why they are doing it and to help them develop conceptual understanding.

Pedagogical content knowledge goes beyond content knowledge and pedagogical skills—it involves integrating knowledge of science, learning, pedagogy, and students and applying that knowledge to science teaching. Pedagogical content knowledge also includes finding ways to make the subject matter comprehensible to others, understanding what makes learning topics easy or difficult, and understanding what conceptions and preconceptions students may have and knowing strategies that will help them reorganize their understandings (Shulman, 1986).

For reform to take place, "professional development must include experiences that engage prospective and practicing teachers in active learning that builds their knowledge, understanding, and ability" (NRC, 1996). This chapter explores this challenge and the issues associated with the preparation and continuing education of chemistry teachers and provides examples of best practices.

Preservice Teachers

Less emphasis on: Separation of science and teaching knowledge
More emphasis on: Integration of science and teaching knowledge

The job of educators is not simply to impart what we know to our students, but rather to engage students in constructing their own understanding and fitting this understanding into their existing framework of knowledge (NRC, 1996). The means to achieve this goal include student-centered practices such as authentic experiential learning—opportunities for students to be involved in the real work of science, thereby developing critical thinking and problem-solving skills.

A disconnect exists in preservice teacher education that weakens the efforts to effectively prepare teachers (Ball, 2000) to lead student-centered classrooms. Preservice science teachers take courses in their disciplines, liberal arts, and education. Unfortunately, a lack of integration among these experiences dilutes the impact of this education. Although education courses challenge prospective teachers to learn the theory and practice of education, the time devoted to practice and integration of the pedagogy into disciplinary content is typically not enough to allow mastery. This problem is exacerbated by the fact that the research scientists who teach many undergraduate science content courses generally focus on traditional methods of teaching, such as lectures and verification laboratory experiences, rather than modeling the research-based pedagogical approaches that preservice teachers are learning about.

Such a disconnect between theory and practice makes it difficult for prospective teachers to acquire the important ability to blend content and pedagogy. If preservice teachers learn science as a collection of facts in which large chunks of information must be digested, then it would seem more likely that they, in turn, would teach this way. Ball (2000) states that the fragmentation of subject matter and pedagogy "leaves teachers on their own with the challenge of integrating subject matter knowledge and pedagogy in the contexts of their work." Furthermore, if these preservice teachers do not master reasoning and understanding in their content area, they may subsequently pass this "disability" on to their students (Arons, 1984). It is crucial that preservice teachers are taught science in ways that model research-based pedagogical approaches and that they have repeated opportunities to practice these approaches.

Preservice teachers must go beyond simply garnering factual knowledge and memorization—they must gain an understanding of how and why things happen, why science is important, how it relates to other ideas, and above all, see the big ideas of science (Shulman, 1986). Even if effective pedagogical methods are implemented, in the absence of adequate teacher content knowledge, student learning and achievement will be compromised, and in some cases, may result in reinforcement of student misconceptions (Kruegar, 2001).

While sound conceptual understanding of science is imperative, Duggan-Haas (2004) notes that "…content understanding without pedagogical skill is fruitless." Preservice teachers are faced with the often overwhelming challenge of learning pedagogy when they have no basis for understanding it until they have tried to implement it. Shulman (1986) poses a question that is central to the challenge of educating science teachers: "How does the successful college student transform his or her expertise in the subject matter into a form that high school students can comprehend?" The answer to this question lies in mentoring of preservice/novice teachers and practical experience in the classroom.

In-the-field experiences become increasingly important in the later stages of preservice teachers' development through clinical work, student teaching, and research. In a review of research, Luft (2003) found that beginning teachers experience a disparity between their teacher training programs and the actual classroom. Proposed solutions include increased independent instruction during student teaching or longer, more significant experiences in the field. A study by Adams and Krockover (1997) found that new science teachers' perceptions of their preservice program included a belief that field experiences should be increased.

Best Practice: Preservice Teachers

An example of exemplary teacher training, The Shady Hill Teacher Training Course is conducted at Shady Hill School in Cambridge, MA. The school's first director, Katharine Taylor, said in a 1937 speech, "The more you think of teaching, the more you realize that it can never be classified as a science. It is nearer to being an art." The program focuses on learning through experience and intense supervision in the following experiences:

Participants begin by observing the classroom and eventually "solo" under the supervision of the directing teacher.

The directing teacher provides guidance on how to introduce a lesson, the pacing of instruction, the importance of repetition, how to break down concepts into digestible pieces, and how to prepare new units of instruction.

Participants attend seminars on issues such as multiculturalism, child development, student assessment, and equity.

The program not only gives participants a deeply significant experience in the field, but it also allows the directing inservice teachers to grow, learn, and reflect on the effectiveness of their own teaching (Archer, 2002).

While this program is similar in some ways to the student teaching that most preservice teachers experience, the Shady Hill course is different in that the rest of the coursework and activities are intimately connected to this teaching experience.

Inservice Teachers

Less emphasis on: Transmission of teaching knowledge and skills by lectures
More emphasis on: Inquiry into teaching and learning

Best Practice: Inservice Teachers

A year-long professional development program conducted with Indiana high school science teachers (and described and researched by Lotter et al. (2006)) highlights some of the essential qualities for a program to help teachers reform their teaching using inquiry. The program offers a rich spectrum of professional development experiences, including summer workshop and research experiences, three one-day academic-year follow-up workshops, and a program Web site. The summer research experience immersed teachers in a research lab environment (placement based on previously indicated interests), in which they joined ongoing investigations as active participants. The workshop components emphasized a plan to help teachers address students' learning difficulties, including the following:

- Identify a student-learning bottleneck and develop a lesson that would address the bottleneck;

- Develop ways to model the problem-solving process for students;

- Develop motivational techniques for students;

- Create practice opportunities for students where the learning activities utilize thought processes to successfully overcome the learning bottleneck;

- Assess student learning through written assignments, drawings, evaluations, etc.; and

- Share learning experiences with other teachers and develop a model of the professional development process that could be shared with colleagues.

Participants reported an increased understanding of inquiry teaching, as well as renewed confidence in using inquiry methods in the classroom. Teacher ideas of inquiry broadened to encompass ideas that had been emphasized in the program, while they reported discovering a renewed empathy for students who continually face unfamiliar situations or content.

A main challenge of educating preservice teachers involves their lack of experience in the classroom on which to base their understandings of complicated pedagogical content knowledge (NRC, 1996). However, the challenge with inservice teacher professional development can be quite the opposite—inservice teachers have a wealth of experience to draw upon; however, this experience may serve as a barrier to change.

In pursuing professional development, inservice teachers draw upon the resources of higher education, science-rich centers, and the scientific community (NRC, 1996). Also, by conducting action research, being involved in the mentoring process, and participating in research at the scientific level, inservice teachers can enhance their growth and learning. It is crucial to the future of science education in our nation that practicing science teachers be able to identify and participate in quality professional development. As pointed out in the Standards for Professional Development chapter of the *National Science Education Standards* (NRC, 1996), we must provide opportunities that are *less* about individual learning, fragmented, one-shot sessions, staff developers as educators, and *more* about collegial and collaborative learning, long-term coherent plans, and staff developers as facilitators, consultants, and planners.

Loucks-Horsley et al. (2003) have suggested a set of elements that professional development programs should feature, given the needs of adult learners. These features are based on the constructivist learning theory, as well as transformative learning, i.e., learning that produces "changes in deeply held beliefs, knowledge, and habits of practice" (Loucks-Horsley et al., 2003, citing Thompson and Zeuli, 1999). The features include

- opportunities for teachers to make connections between their existing ideas and new ideas;
- opportunities for active engagement, discussion, and reflection to challenge existing ideas and construct new ones;
- activities that challenge teachers' thinking by producing and then resolving dissonance between old and new ideas;
- learning in contexts which teachers find familiar; and
- help for teachers to develop a range of strategies that address student learning across all levels.

These features resonate with the constructivist view of learning, as does the NRC (1996) with their position that "*learning is something students do, not something that is done to them.*" Learners must use their existing repertoire of experiences and ideas as building blocks or bridges to new learning. As a result, Spigner-Littles and Anderson (1999) point out that instructors of professional development for teachers should assume the role of a facilitator or moderator to keep the teachers focused on sharing their previous experiences, views, and perspectives.

Effecting Change

Less emphasis on: Teacher as target of change
More emphasis on: Teacher as source and facilitator of change

The goal of any professional development program for teachers is to effect change (NRC, 1996). Change may occur in teachers' beliefs, attitudes, content knowledge, or other facets, but the end result must be change in classroom practice. Many programs fail to effect change because they do not address the special needs of teachers who are contemplating change.

The Concerns-Based Adoption Model (Hall and Hord, 1987) describes the stages of concerns for teacher-learners who are experiencing change. At each stage of concern, the teacher-learner asks questions. In the early stages, questions are more self-oriented in nature; for example, "what is it?" and "how will it affect me?" After these questions are resolved, questions become more task-oriented, such as "how do I do it?" In the latter stages, individuals focus on the impact of the changes at hand. For example, teachers begin to question whether the new process is working for their students or whether the process can be modified to work better (Loucks-Horsley, 1996). Professional developers can capitalize upon this model by addressing their presentation to each stage of questioning. For example, a teacher may need to learn how a process will affect them (e.g., "how much class time will this actually take?") before they can learn how it will help their students. Depending on what type of change is being effected, different stages of concern will take varying degrees of time. The need for sustained support and follow-up is inherent in this model. The strength of this model is in the idea that individuals and their various needs for information, assistance, and moral support are considered while developing professional development activities (Loucks-Horsley, 1996).

As older learners become attached to their practices, beliefs, knowledge, values, and views, they are likely to reject or explain away new ideas (Spigner-Littles and Anderson, 1999). As a result, professional developers must tackle this issue head on by empowering teachers to buy into the change and support them over time with well-constructed follow-up activities. Sarquis (2001) offers recommendations for continued follow-up and sustained impact in professional development activities that will provide continued motivation and support while sustaining the creative momentum initiated during workshops. Some of these recommendations include hosting return visits for the teachers, supporting teachers' participation in professional meetings, and helping teachers to increase their leadership roles and responsibilities within their districts.

One professional development tool that encourages and supports change in teaching practice is action research. Action research allows teachers to reflect on their teaching practices by analyzing their teaching and their students' learning. Loucks-Horsley et. al. (2003) suggest that that the strength of action

---

Best Practice: Leadership

Professional development programs do not often include development of teacher leadership skills and teacher leaders. One of the exceptional aspects of North Carolina's FIRST (Fund for the Improvement and Reform of Schools and Teaching) programs, as described by Wallace et al. (2001), was the inclusion of strategies to develop teacher leaders and support them in their new roles to bring about reform of their entire schools and not just individual classrooms. In order to ensure that these professional development programs were effective, a series of four critical elements and accompanying strategies were put into place (Wallace et al., 2001):

- Designing and implementing long-term professional development by

  o building a multiphased 18-month program that included preassessment sessions, summer institute, academic year follow-up sessions, and a final workshop the following summer and

  o creating time for reflection.

- Building teachers' capacity for shared decision making by

  o administering a needs assessment to help identify strengths and weaknesses,

  o designing a school improvement plan (SIP) with objectives and strategies for addressing their needs,

  o allowing for review of professional development activities that modeled the shared decision-making process and ensured that the activities would assist the teacher leaders in their SIPs, and

  o processing leadership content and practicing leadership skills.

- Creating a supportive environment for the teacher leaders by

  o providing principal support, including resources, release time, space, encouragement, and praise,

  o allowing time for two-teacher team collaboration, which provided the teachers with support from each other during the school year,

  o including time for teams of teacher leaders to problem solve with other teams, and

  o including project staff support during implementation phase, including school site visits, telephone calls, workshop presentations at the school, and one-on-one discussions.

- Incorporating assessments by conducting

  o formative assessment and

  o summative evaluation.

research as a form of professional development is that the teachers themselves are developing the research questions or are contributing to the questions in a meaningful way. This gives them ownership and control over their learning; thus, they are committed to promoting changes in their teaching practices as revealed through their research results.

Professional development that effects change in teacher practice often leads to the professional growth of teachers as they assume increased leadership and responsibility within their school districts (Sarquis, 2001). Teachers who move up the ranks of the leadership development path develop stronger content and pedagogical knowledge, work toward development and implementation of materials and curriculum in the classroom, and also work toward outreach to other teachers and the community, thereby synergizing the change they have undergone to impact their colleagues.

Professional development can also empower teachers to become leaders in their schools, districts, and professional organizations. Such leadership enables teachers to become powerful agents of change in their own school reform. A study by St. John and Pratt (1997) (as cited in Pratt, 2001) pointed to leadership as an essential factor for reform. The study suggests that the best cases of school reform occurred where there was long-term, committed leadership. Leaders who were connected to many sources of support and focused primarily on educational substance were critical to change, as were leaders who use the standards and policy as vision to guide their reform.

Issues of Retention

Less emphasis on: Teacher as an individual based in a classroom
More emphasis on: Teacher as a member of a collegial professional community

The issue of retention of teachers in the profession goes hand in hand with the topic of professional development (NRC, 1996). Simply put, effective professional development can reduce teacher turnover rates (Smith and Rowley, 2005), which are relatively high in comparison to those of other types of employees. Beginning teachers have a very high turnover rate, while rates decline during midcareer years and rise again toward retirement. On average, about 29% of all new teachers leave the profession after three years (Ingersoll, 2003).

In an effort to retain teachers, many new initiatives provide support for beginning teachers. Examples of support that has been implemented within these initiatives include focusing on practical issues like classroom management and discipline, online discussions with professors of education, mentoring, peer assistance, and other forms of guidance and support (DePaul, 2000). The U.S. Department of Education's Survival Guide for New Teachers also lists ways in which beginning teachers can effectively collaborate with veteran teachers, parents, principals, and college and university faculty to obtain the support that they need during the first few critical years of teaching.

Best Practice: Retention

The Valle Imperial Project in Science (VIPS) is a professional development plan for teacher retention and renewal in California (Klentschy et al., 2003). Teacher retention and renewal, as well as development of a broad countywide base of sustained teacher leadership, are the goals of this project. Briefly summarized, the 10 elements of this action plan are

- developing a link between preservice and actual classroom practices through specific science methods courses that are linked to teacher practices,

- local and national institutes and opportunities to attend a local master's degree program to deepen teacher content understanding,

- opportunities for teachers to strengthen their pedagogical skills,

- in-classroom support from science resource teachers, including weekly visits to novice teachers,

- leadership development of teachers, where lead teachers act as leaders in professional development activities and work to bring about systemic change,

- materials support, where teachers receive VIPS materials on a quarterly basis,

- time for collaboration and networking within and between schools,

- applications of technology, including workshops on technology applications for teachers as well as online professional development activities where teachers can communicate with each other,

- workshops focusing on student work as the centerpiece of standards-based performance assessment, and

- opportunities to refine instructional delivery through reflection and lesson study groups, where there is a cyclic process of lessons being taught, observed, and refined for improvement.

Teacher collaboration and mentoring are important aspects of teacher retention. Mentoring is a form of professional development in which a new teacher and an experienced teacher (or sometimes a scientist) collaborate to provide support for the new teacher and enhance the leadership role of the mentor (Loucks-Horsley et al., 2003). Experienced teachers provide new teachers with emotional support. However, in order to utilize mentoring in its fullest capacity, mentors have many other responsibilities. Some of the responsibilities of a mentor toward the new teacher, as pointed out by Dunne and Newton (2003), include enabling new teachers to increase their knowledge of content, guiding them toward ways of including scientific inquiry into teaching, and sharing ideas on making the subject matter comprehensible to others. Mentoring experiences also promote professional development and growth for experienced teachers. Through mentoring, experienced teachers are rejuvenated and renewed by the newer teachers as they reexamine and reflect upon their own teaching practices (Archer, 2002; Brennan, 2003).

Self-efficacy is another promising tool to improve teacher retention. If teachers believe that they have the ability both to present knowledge and to guide students toward learning, then their job satisfaction is likely to increase. In a study by Khourey-Bowers and Simonis (2004), it was found that following professional development programs that focused upon enhancing self-efficacy, there was a significant increase in participants' self-efficacy and that their beliefs significantly improved in relation to their ability to effectively teach science concepts. Additionally, Louis (1998) found that

> "Enhanced opportunities within the school to use and develop new skills—one of the most important predictors of both efficacy and commitment—was most effectively promoted by all-school, teacher organized inservice activities, teacher mentoring programs, or programs that provided grants to teachers to develop new programs."

Special Challenges of Second-Career Teachers

Less emphasis on: Teacher as a technician
More emphasis on: Teacher as intellectual, reflective practitioner

Transitioning from one career to the next involves special challenges, particularly concerning the transition into teaching. Scharberg (2005) suggests that for those who are considering making the transition to teaching, a recommended first step is for the decision maker to decide whether he or she has the qualities to become a successful chemistry teacher: love of teaching adolescents, solid knowledge of chemistry, being a team player, excellent communication and organizational skills, and a large supply of patience (NRC, 2006).

The transition is difficult, because most prospective teachers in this category have undergraduate and graduate degrees in the content area; however, they do not have knowledge of instructional strategies or pedagogical content knowledge. Gerald Wheeler, National Science Teachers Association executive director, feels that most scientists cannot just step into the classroom. He believes "it's imperative that the scientist-turned-teacher have a very good handle on the learning characteristics of the students [with whom] they're going to be interacting." However, because these professionals are familiar with the use of inquiry-based methods from their

Best Practice: Second Careers

The Teacher Recruitment and Induction Project (TRIP) is a best-practice program specifically developed for midcareer adults to make the transition into teaching (Resta et al., 2001). Intended for those who have already received their bachelor's degrees, the program allows participants to complete initial certification requirements in one year or less. TRIP focuses on providing participants with strong content and pedagogical knowledge through integrated coursework where instructors model best teaching and instructional practices. In addition to first-semester coursework, participants also spend 2 days per week in the field, observing, tutoring, and teaching students. Participants' field experiences are integrated into coursework. During the second semester, participants become student teachers, complete one more graduate course, and meet weekly with a mentor. After initial certification, the program offers intensive induction support via the mentors during their first two years of teaching.

first profession, they may "be able to take a child into an authentic inquiry exercise" (Henry, 2003). Vicki A. Jacobs, associate director of Harvard's Teacher Education Program, also believes that those making a midcareer transition to teaching will have more life applications and knowledge to draw upon (Henry, 2003).

Recommendations and Conclusions

The nature of science is that it does not stand still; science is ever changing and growing as new theories are formed, challenged, and then either accepted or refuted. Likewise, the nature of professional development for chemistry teachers (and all science teachers) must follow a similar path of growth and discovery, just as teachers themselves must continually adapt and change to meet the needs of their students. Professional development—in all its forms—is perhaps the most important tool in our arsenal for helping and supporting teachers as they shape our future scientists, voters, and decision makers.

As we continue to discover more about the science of the brain and how people learn, our work with teachers will evolve correspondingly. New and emerging technologies will provide new avenues for impacting teachers. Virtual professional development is a significant option for the future. However, most current online offerings do not represent what the future can be: they are passive rather than interactive. True virtual professional development will include high-level problem solving, decision making, interpretation, and analysis through partial or full simulations (Charles and Griffin, 2007).

Professional developers of chemistry teachers are in the position to play a substantive role in shaping this future. A periodic check of your methods against best practices in the field (both new and well established) will help you to best play this role.

Through a combined 70 years of experience in science education, we have established a model for teacher professional development. This protocol has been recognized by the U.S. Department of Education as a "State Model Program," by the National Science Foundation Project Kaleidoscope as a "Program that Works," and by the Chemical Manufacturers Association as a "Recommended Model Program." Ask yourself how well your efforts parallel the best practices in professional development for chemistry teachers shown in Table 1. (These practices are paired with corresponding elements from the Standards for Professional Development for Teachers of Science chapter in the *National Science Education Standards*.)

Table 1: Best Practices in Professional Development for Chemistry Teachers

Program Administration Teachers, administrators, and scientists are included in program development and implementation.	
Less emphasis on	*More emphasis on*
Reliance on external expertise	Mix of internal and external expertise
Teacher as consumer of knowledge about teaching	Teacher as producer of knowledge about teaching
Vision for the Classroom Emphasize a hands-on, minds-on instructional approach; a balance between content and process in classroom instruction; active assessment of important learning outcomes; and the use of materials, strategies, and perspectives sensitive to diversity.	
Less emphasis on	*More emphasis on*
Transmission of teaching knowledge and skills by lectures	Inquiry into teaching and learning
Learning science by lecture and reading	Learning science through investigation and inquiry

Separation of science and teaching knowledge	Integration of science and teaching knowledge
Separation of theory and practice	Integration of theory and practice in school settings

Teacher Development Program Activities

Offer a range of instructional strategies, including activity-based instruction, guided and open inquiry, scenario-based investigations, and learning cycles. Model the teaching practices intended to be transferred to the classroom. Engage teachers in activities that provide a foundation for them to actively construct their own knowledge. Use the tools, methods, processes, and real-world challenges of science. Give teachers opportunities to plan ways to incorporate new information into their curricula.

Less emphasis on	More emphasis on
Courses and workshops	Variety of professional development activities

Contribution of Partnering Scientists

Collaborate with scientists from industry, government, and the private sector to provide valuable and complementary perspectives to professional development instruction.

Less emphasis on	More emphasis on
Teacher as an individual based in a classroom	Teacher as a member of a collegial professional community

Follow-Up

Include both formal and informal follow-up mechanisms both to sustain the creative momentum generated in workshops and to promote the understanding and ability for lifelong learning advocated by the national standards.

Less emphasis on	More emphasis on
Fragmented, one-shot sessions	Long-term coherent plans

Teacher Leadership and Responsibility

Empower graduates to assume leadership roles in effecting systemic change within their districts and schools. Involve district teams of teachers and administrators and require the development of action plans to implement programming.

Less emphasis on	More emphasis on
Individual learning	Collegial and collaborative learning
Staff developers as educators	Staff developers as facilitators, consultants, and planners
Teacher as technician	Teacher as intellectual, reflective practitioner
Teacher as follower	Teacher as leader
Teacher as target of change	Teacher as source and facilitator of change

Program Evaluation

Include frequent evaluation of professional development efforts in your plans. Use both quantitative and qualitative measures to assess the effects of your program on participant knowledge and attitudes toward science, as well as the effects of the program on students, systemic changes, and second-tier outreach.

ACS Education Division

Above all, make sure that your professional development efforts involve modeling the pedagogical approaches you hope to impart to teachers. Your successful efforts will lead teachers to more positively impact their classrooms for years to come.

Recommended Readings

Sarquis, A. M. Recommendations for Offering Successful Professional Development Programs for Teachers. *J. Chem. Educ.* 2001, *78*, 820–823. This article shares insights gained by administering three major professional development initiatives. As Director of the Miami University Center for Chemistry Education (CCE), Sarquis developed the CCE teacher professional development model.

Professional Development Leadership and the Diverse Learner; Rhoton, J., Bowers, P., Eds.; Issues in Science Education; NSTA Press: Arlington, VA, 2001. This book presents chapters that effectively address the issues of professional development. Each chapter illustrates the utility of professional development for practitioners and addresses general issues and perspectives related to science education reform. The examples provided are valuable models for those conducting professional development activities.

Recommended Web Sites

Triangle Coalition. http://www.trianglecoalition.org (accessed June 30, 2008). The Triangle Coalition is a Washington, D.C.-based nonprofit organization comprising members from business, education, and scientific and engineering societies. Triangle Coalition's Mission is to bring together the voices of these stakeholders to improve the quality and outcome of mathematics, science, and technology education. Triangle Coalition focuses its action in advocacy, communication, and programmatic efforts to advance science, mathematics, and technology education for all students. The Web site provides valuable updates for all who are interested in the state of science education in the United States, as well as news about upcoming professional development opportunities.

The TE-MAT Project. http://www.te-mat.org/ (accessed June 30, 2008). The TE-MAT (Teacher Education Materials) Project supports professional development providers, as they work to enhance the capacity of preservice and inservice teachers to provide high-quality K–12 mathematics/science education. Current national standards for mathematics and science education are based on the premise that K–12 education should provide powerful mathematics and science content for all students and should focus on teaching for understanding. This vision served as a foundation for the design of the TE-MAT database. The site is designed for all levels of experience—from novice to expert—to help in the design, implementation, and evaluation of effective professional development programs to support teachers.

References

Adams, P. E; Krockover, G. H. Concerns and Perceptions of Beginning Secondary Science and Mathematics Teachers. *Sci. Educ.* 1997, *81*, 29–50.

American Chemical Society. Science Education Policies for Sustainable Reform. http://www.chemistry.org/portal/resources/?id=c373e90165f4c0448f6a17245d830100 (accessed Feb. 2007).

Archer, J. Tools of the Trade. *Education Week* 2002, *21,* 30–35.

Arons, A. B. Education Through Science. *J. Coll. Sci. Teaching* 1984, *13,* 210–220.

Ball, D. L. Bridging Practices: Intertwining Content and Pedagogy in Teaching and Learning to Teach. *J. Teacher Educ.* 2000, *51,* 241–247.

Brennan, S. Mentoring for Professional Renewal: The Kentucky Experience. In *Science Teacher Retention: Mentoring and Renewal,* Rhoton, J., Bowers, P., Eds.; Issues in Science Education; NSTA Press: Arlington, VA, 2003; pp 161–169.

Charles, K. J.; Griffin, J. E. Teaching Science in the 21st Century: Virtual Professional Development: The Good, the Bad, and the Future. *NSTA Rep.* Feb 5, 2007. http://www3.nsta.org/main/news/stories/nsta_story.php?news_story_ID=53347.+ (accessed February 2007).

DePaul, A. *Survival Guide for New Teachers: How New Teachers Can Work Effectively With Veteran Teachers, Parents, Principals, and Teacher Educators;* U.S. Department of Education: Office of Educational Research and Improvement, 2000.

Duggan-Haas, D. A Proposed Introduction to the NSTA Standards for Science Teacher Preparation. http://www.msu.edu/~dugganha/intro.htm (accessed May 2004).

Dunne, K. A.; Newton, A. Mentoring and Coaching for Teachers of Science: Enhancing Professional Culture. In *Science Teacher Retention: Mentoring and Renewal,* Rhoton, J., Bowers, P., Eds.; Issues in Science Education; NSTA Press: Arlington, VA, 2003; pp 71–84.

Dyasi, H. M.; Dyasi, R. E. "Reading the World Before Reading the Word": Implications for Professional Development of Teachers of Science. In *Crossing Borders in Literacy and Science Instruction: Perspectives on Theory and Practice,* Saul, E. W., Ed.; NSTA Press: Arlington, VA, 2004; pp 420–446.

Hall, G. E.; Hord, S. M. *Change in Schools: Facilitating the Process;* SUNY Series in Education Leadership; State University of New York Press: Albany, NY, 1987.

Henry, C. M. Back to School: High School Teaching Attracts Both Younger and Midcareer Chemists with Advanced Degrees. *Chem. Eng. News 81,* 47–49.

Ingersoll, R. M. Turnover and Shortages Among Science and Mathematics Teachers in the United States. In *Science Teacher Retention: Mentoring and Renewal,* Rhoton, J., Bowers, P., Eds.; Issues in Science Education; NSTA Press: Arlington, VA, 2003; pp 1–12.

Khourey-Bowers, C.; Simonis, D. G. Longitudinal Study of Middle Grades Chemistry Professional Development: Enhancement of Personal Science Teaching Self-Efficacy and Outcome Expectancy. *J. Sci. Teacher Ed.* 2004, *15,* 175–195.

Klentschy, M. P.; Molina-De La Torre, E. A Systemic Approach to Support Teacher Retention and Renewal. In *Science Teacher Retention: Mentoring and Renewal,* Rhoton, J., Bowers, P., Eds.; Issues in Science Education; NSTA Press: Arlington, VA, 2003; pp 161–169.

EDThoughts: What We Know About Science Teaching and Learning; Kruegar, A., Sutton, J., Eds.; McREL: Aurora, CO, 2001.

Lotter, C.; Harwood, W.; Bonner, J. Overcoming a Learning Bottleneck: Inquiry Professional Development for Secondary Science Teachers. *J. Sci. Teacher Ed.* 2006, *17,* 185–216.

Loucks-Horsley, S. Professional Development for Science Education: A Critical and Immediate Challenge. In *National Standards & the Science Curriculum,* Bybee, R., Ed.; Kendall/Hunt Publishing: Dubuque, IA, 1996.

Loucks-Horsley, S.; Stiles, K.; Hewson, P. Principles of Effective Professional Development for Mathematics and Science Education: A Synthesis of Standards. *NISE Brief* 1996, *1,* 1–6.

ACS Education Division

Loucks-Horsley, S.; Love, N.; Stiles, K. E.; Mundry, S.; Hewson, P. W. *Designing Professional Development for Teachers of Science and Mathematics,* 2nd ed.; Corwin Press: Thousand Oaks, CA, 2003.

Louis, K. S. Effects of Teacher Quality of Work Life in Secondary Schools on Commitment and Sense of Efficacy. *School Effectiveness and School Improvement* 1998, *9*, 1–27.

Luft, J. A. Induction Programs for Science Teachers: What the Research Says. In *Science Teacher Retention: Mentoring and Renewal,* Rhoton, J., Bowers, P., Eds.; Issues in Science Education; NSTA Press: Arlington, VA, 2003; p 35–44.

National Research Council. *National Science Education Standards;* National Academy Press: Washington, DC, 1996.

Pratt, H. The Role of the Science Leader in Implementing Standards-Based Science Programs. In *Professional Development Leadership and the Diverse Learner,* Rhoton, J., Bowers, P., Eds.; Issues in Science Education; NSTA Press: Arlington, VA, 2001; pp 1–9.

Resta, V.; Huling, L.; Rainwater, N. Preparing Second-Career Teachers. *Ed Leadership* 2001, *58*, 60–63.

Sarquis, A. M. Recommendations for Offering Successful Professional Development Programs for Teachers. *J. Chem. Educ.* 2001, *78*, 820–823.

Scharberg, M. A. Teaching High School Chemistry as a Second Career. *J. Chem. Ed.* 2005, *82*, 1281–1285.

Shulman, L. S. Those Who Understand: Knowledge Growth in Teaching. *Ed. Researcher* 1986, *15*, 4–14.

Smith, T. M.; Rowley, K. J. Enhancing Commitment or Tightening Control: The Function of Teacher Professional Development in an Era of Accountability. *Educ. Pol.* 2005, *19*, 126–154.

Spigner-Littles, D.; Anderson, C. E. Constructivism: A Paradigm for Older Learners. *Educational Gerontol.* 1999, *25*, 203–209.

St. John, M.; Pratt, H. The factors that contribute to the "best cases" of standards-based reform. *School Sci. Math.* 1997, *97*, 316–324.

Thompson, C. L.; Zeuli, J. S. The frame and the tapestry: Standards-based reform and professional development. In *Teaching as the Learning Profession: Handbook of Policy and Practice,* Darling-Hammond, L.; Sykes, G., Eds.; Jossey Bass: San Francisco, CA, 1999; pp 341–375.

U.S. Department of Education. The Mission and Principles of Professional Development. http://www.ed.gov/G2K/bridge.html (accessed Feb. 2007).

Wallace, J. D.; Nesbit, C. R.; Newman, C. R. Bringing About School Change: Professional Development for Teacher Leaders. In *Professional Development Leadership and the Diverse Learner,* Rhoton, J., Bowers, P., Eds.; Issues in Science Education; NSTA Press: Arlington, VA, 2001; pp 37–47.

Chapter 11

Assessment of Student Learning

by Laura Slocum and Thomas A. Holme

Thomas Holme *received his Ph.D. from Rice University and is a professor of chemistry at the Iowa State University. He has been the Director of the Examinations Institute of the Division of Chemical Education of American Chemical Society since 2002 and conducts research on how to measure student learning in chemistry as a key component of his scholarship. Contact e-mail: taholme@iastate.edu*

Laura Slocum *graduated from Western Connecticut State University with a B.A. in chemistry and from Ball State University with an M.S. in chemistry. She has taught high school chemistry for the past 16 years in both Connecticut and Indiana. Presently, Laura teaches at University High School of Indiana. She serves as the Secondary School Associate Editor for the* Journal of Chemical Education. *She also serves on the Board for Trustees for the Examinations Institute and as an Alternate Councilor for the Division of Chemical Education. Contact e-mail: lslocum@universityhighschool.org*

Introduction

The importance of assessment in any discipline at any level is nearly self-evident. Ultimately, for pragmatic reasons alone, the nature of assessments plays a major role in the practical formulation of what occurs in the classroom; if the assessments for a course are aligned to well-defined state or national standards and if these same assessments require students to acquire deep understanding, then both students and teachers will respond. More recently, the emphasis on assessment within the No Child Left Behind (NCLB) Act has provided even more motivation for teachers to understand how to better assess their students, particularly, in line with state and national standards.

Nurrenbern and Pickering (1987) made a rather startling observation that "the last 30 years of chemistry can be characterized by the 100 most popular questions." The two decades since this statement have seen important developments in assessment (Pellegrino et al., 2001) to be sure, but a glance at any textbook test bank reveals many of the same questions that constituted chemistry tests in the 1980s are still present in today's exams. Although this longevity may simply speak to the veracity of the fundamental knowledge of chemistry, it also suggests a more careful inspection is warranted regarding the assessment of learning in chemistry. What are the possibilities for constructing chemistry assessments? How do these assessment techniques play a critical role in the development of the teaching of chemistry at all levels?

Significant effort has resulted in the development of assessments related to chemistry in many states. For example, the state of Indiana used a document that detailed student learning objectives known as "Chemistry Proficiencies" prior to the adoption of state academic standards in 2000. On the basis of these proficiencies, the Indiana Department of Education (DOE) and a small group of teachers prepared a test called a Core 40 Exam. This exam was available for chemistry teachers to use, voluntarily, at the end of the school year to test student proficiencies in chemistry. Once the state standards were established in 2000, the DOE started working with various groups of teachers and the Center for Innovation in Assessment at Indiana University to prepare a chemistry test that mapped directly to state standards. This process went through several iterations and in May 2005, the Core 40 Classroom Assessment–Chemistry I was published. This exam has two sections, containing mostly multiple-choice questions with some open-ended response questions at the end of each section. Administration of this exam is not required, but the exam is available online at a password-protected site for teachers to use. Because this exam is not required, no statistical information is collected on student performance. There is presently no required exam in chemistry in Indiana; however, starting in the 2007/2008 school year, there is a required Biology I–End-of-Course Assessment that all students must take. Other states have undertaken similar processes.

Chemistry, as a discipline, also has the distinction of a long-standing effort for nationally normed exams as produced by the Examinations Institute of the American Chemical Society (ACS). The first high school exam was released in 1954, and new exams are produced every two years, as is described more fully below under "Available Resources." Therefore, there exists a range of concepts and guidelines that can be utilized to enhance assessment efforts in high school chemistry. In the remainder of this chapter, we describe important terminology for assessment and survey resources for assessment that are constructed with national standards in mind. We describe in detail how ACS exams are constructed and how teacher involvement in the process leads to important professional development activities. Lastly, we conclude by discussing how the principles of assessment can be incorporated into teacher-designed test items.

Constructing Assessments

Individuals who study to become chemistry teachers typically spend most of their time developing their content knowledge, while comparatively little time is spent studying measurement of student learning via tests. The scholarship associated with developing robust tests (and the questions on these tests) remains an area of active research, but some guiding principles have been established over the years. Regardless of the identity of the test generator—a national society, the state department of education, or an individual classroom teacher—there are two key ideas that provide the fundamental framework needed to assess educational measurement efforts, namely validity and reliability (Popham, 2003).

Table 1: Face Validity of Chemistry Questions

Less face validity	More face validity
1. How long is 4.7 yards in mm? 2. Which process is exothermic, one that has $\Delta H = -315$ kJ or one with $\Delta H = 356$ kJ?	1. A chemical bond is 124 pm long; express this distance in meters. 2. Which energy change could represent a combustion reaction, one with $\Delta H = -315$ kJ or one with $\Delta H = 356$ kJ?

Validity is best considered as a measure of whether or not a test measures what we intend it to measure. There are several categories of validity. The first, and perhaps most apparent form, is referred to as *face validity*. At this level, the question that one might ask of a chemist or chemistry instructor is, "Is the item being considered a chemistry question?" This question may seem almost too obvious to ask, but consider the role of background knowledge, particularly in mathematics, for a chemistry course. It is common for a chemistry teacher to ask questions related to dimensional analysis that are not related to chemistry, such as the conversion of a distance in an obscure unit system (furlongs, for example) into metric distances. Thus, considering face validity, Table 1 suggests two ways that teachers consider asking similar questions, but one way has more face validity and the other less.

The next important category for validity is called *content validity*, and it is a measure of whether the item being considered actually measures the content it is expected to answer. Accuracy of content is incorporated in this concept, but it is richer than accuracy alone. Content validity addresses the question of whether or not a test assesses the skills and knowledge of the course for which it is intended. In Table 2, we present two sample questions related to NSES Standard B1: "*All students should develop an understanding of structure of atoms, structure, and properties of matter... .*" Each sample question is worded in two different ways to illustrate how a question can have more content validity and less content validity.

Table 2: Content Validity of Chemistry Questions Related to NSES Standard B1

Less face validity	More face validity
1. What is the atomic number of Bromine, Br? 2. What is the atom with an electron configuration of $1s^2 2s^2 2p^6 3s^1$?	1. Describe how the atomic number and group location of an element on the periodic table provides information about the element? 2. Describe how the electron configurations of sodium and potassium are similar.

Finally, a more technical component of test writing is called *construct validity*, which measures the extent to which the format, language, and other factors of the question influence the measurement of knowledge. Many questions provide unintentional hints about correct answers (or ways to eliminate some of the incorrect answers, i.e., distractors.) In many cases, a form of statistical analysis, called item statistics, is capable of providing insight into the construct validity, particularly for multiple-choice items.

The two most important concepts related to item statistics are the difficulty and discrimination associated with each question. The difficulty of a question is traditionally defined as the fraction of students who answer it correctly. A high-difficulty index means a large fraction of the students who take the exam answer the question correctly, so it is a relatively "easy" question. Discrimination measures the relative performance on a question between high-proficiency students and low-proficiency students. It is traditionally calculated by looking at the top 25% of students (defined by total score on the overall exam) and the bottom 25% of students. Discrimination is the fraction correct in the top 25% minus the fraction correct in the bottom 25%.

For norm-referenced exams, the difficulty of an individual item will typically vary from 0.4 to 0.7, so that there is an ability to spread out overall scores in a sample of students. Discrimination is also important for norm-referenced exams because the premise of a norm-referenced exam is that student scores are spread out to discriminate those who know the content from those who do not.

This discussion has focused so far on validity at the item level. Another key idea is reliability, which is a measure of whether a test would measure the same performance if a student were to retake it. Because students seldom take the same test twice, statistical estimates are used for reliability. These estimates often are akin to comparing how students perform on the first half of the test versus their performance on the second half of the test.

For teachers who face new expectations for assessment based on NCLB or other factors, these definitions can help them decide what must be considered in creating or choosing assessment materials. We will now look at some resources that are available and then consider teacher-developed assessment materials.

Available Resources

Plenty of potential resources are available to high school chemistry teachers in terms of assessment. Quantity is not a problem. Rather, the key challenge lies in determining how well these resources align with national standards. The Examinations Institute of the American Chemical Society produces nationally normed exams for both first-year high school chemistry and advanced high school chemistry. The first year exam is released every other year (in odd years), and, since 2003, has been designed with the goal of matching the content standards for chemistry present in the *National Science Education Standards* (NSES). Table 3 shows the distribution of items associated with the pertinent standards for the three most recently released exams.

Table 3. Alignment of Items From ACS First-Year High School Chemistry Exams with NSES Standards

NSES Standard	Standard Descriptions: "Students should develop …"	Number of questions (HS03)*	(HS05)	(HS07)
A1	… abilities to do scientific inquiry.	11	6	6
A2	… an understanding about scientific reasoning.	33	1	1
B1	… an understanding of the structure of atoms.	5	7	8
B2	… an understanding of the structure and properties of matter.	31	44	38
B3	… an understanding of chemical reactions.	28	15	23
B5	… an understanding of the conservation of energy and increase in disorder.	10	4	4
B6	… an understanding of interactions of energy and matter.	0	3	2
D1	… an understanding of energy in the earth system.	1	0	0
F5	… an understanding of natural and human-induced hazards.	1	0	0

*Note that HS03 refers to the 2003 First Year High School Chemistry Exam, HS05 to 2005, and HS07 to 2007.

Table 3 would appear to suggest that there were more questions on the 2003 exam, than in 2005 or 2007. In 2003, however, questions that the test-writers believed matched to multiple standards were double or even triple counted in some cases. Beginning in 2005, the directions given to test-writers were changed so that the questions they submitted would map to only one national standard.

With regard to national assessment programs, perhaps the College Board and the Advanced Placement (AP) tests represent the most widely utilized resource for chemistry teachers. The College Board offers a Subject Test (formerly SAT II: Subject Test) in Chemistry. This test is designed for students that have taken a one-year college preparatory chemistry course. It is also important to note that the College Board is currently engaged in an extensive, NSF-funded process to design a large-scale change in all AP science courses, including chemistry.

Alternative forms of assessment have also been devised for the chemistry classroom. The use of portfolios, for example, has steadily developed over the past decade (Phelps, 1997). At the high school level, the International Baccalaureate (IB) program uses a portfolio approach, particularly related to laboratory work (IBWeb, 2007).

Assessment and Professional Development

Few teachers have access to professional development opportunities related to assessment from within their content-based specialty. For example, few workshops in writing good test questions in chemistry are offered at national conferences related to chemical education.

The ACS Division of Chemical Education (DivCHED) has sponsored the Examinations Institute for many years and provides opportunities for assessment-based professional development. For example, workshops at summer conferences allow teachers to experience, in an abbreviated format, the process by which an ACS exam is written by a test committee. When an ACS Exam is constructed, the process involves several important steps that help teachers at all levels become more versed in matters related to the construction of test questions.

Construction of the first-year high school exam begins two years prior to the release of the exam by a committee of chemistry teachers. Each exam committee is composed of approximately 20 members—4 or 5 members have served on two or more previous committees, 5–10 members have served on one prior test committee, and there are typically 4 or 5 new committee members. Each committee member is assigned specific topics and provided writing style directions for the test questions they are asked to submit for possible use on the final exam. Each question is mapped to an NSES indicator by the initial author of the question. Each member submits at least five questions for each assigned topic area, so that the committee begins its work with about 450 potential questions.

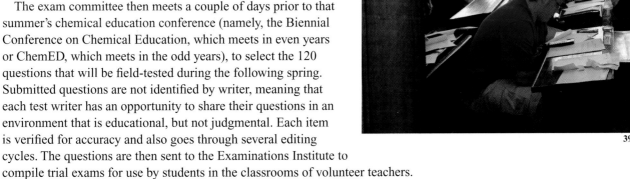

39th IChO

The exam committee then meets a couple of days prior to that summer's chemical education conference (namely, the Biennial Conference on Chemical Education, which meets in even years or ChemED, which meets in the odd years), to select the 120 questions that will be field-tested during the following spring. Submitted questions are not identified by writer, meaning that each test writer has an opportunity to share their questions in an environment that is educational, but not judgmental. Each item is verified for accuracy and also goes through several editing cycles. The questions are then sent to the Examinations Institute to compile trial exams for use by students in the classrooms of volunteer teachers.

In the spring of the second year of each exam development cycle, the 120 questions are field-tested, and the Examinations Institute calculates the item statistics for each question. The committee members use these statistics to identify items that perform well (per the criteria discussed earlier in this chapter, such as difficulty and discrimination). During the summer chemical education conference of the second year, the committee selects the final 80 questions for the newest first-year high school exam.

Test committee members have consistently noted that serving on an ACS exam committee is one of their most rewarding and professionally enriching experiences. They find it not only educationally informative, but also personally rewarding. Committee members become great support colleagues for each other and continue to learn from each other throughout their careers.

Teacher–Designed Assessment

Although nationally normed tests provide important forms of assessment in the current educational climate, most assessment remains teacher designed. Several key concepts that tend to enhance the quality of test questions are worth noting here.

First, assessment related to standards must reflect the emphases of the standards. Within most systems, this requirement will mean questions that move away from easily measured, discrete knowledge, toward items that align with the expectations of the national (and state) standards. In particular, teachers have expressed a greater desire and need for inquiry-based or reasoning questions. This does not necessarily require open-ended questions or questions about specific laboratory experiments. For example, adding a laboratory context (as shown in Fig. 1B), will elicit greater reasoning skills within a multiple-choice item.

Figure 1. Examples of Items

What is the pH of a 0.00001 M HNO_3 solution?		If an experiment requires a pH of 5, which solution could be used?	
(A)	1	(A)	0.5 M HNO_3
(B)	3	(B)	0.5 M CH_3COOH
(C)	5	(C)	0.00001 M HNO_3
(D)	9	(D)	0.00001 M CH_3COOH

Second, the objectives for a specific test should be formulated in terms of words that require student action. Rather than starting a question that begins with "What is …" terms such as demonstrate, describe, distinguish, or assess provide a richer environment for devising test questions. This difference is particularly important with open-ended questions. Fig. 2 shows how such wording changes can produce a more robust level of student response for similar chemical content.

Figure 2. Examples of Items

What is the precipitate when $AgNO_3$ and HCl are mixed?	Describe how to distinguish between a colorless solution that contains silver ions and one that does not.

Third, when writing items, every effort should be made to ensure that each item (1) be written concisely and with minimal complications; (2) avoid adding irrelevant material, recognizing that if the objective of the item is to assess student analysis of data, it is possible to have more data than needed, but this would not qualify as irrelevant information; (3) the problem being asked should be clear enough that the student is not confused about what is expected; (4) avoid long descriptions about what is expected; and (5) avoid phrasing the question in negative terms. Multiple-choice items add additional constraints, in particular, the structure of the responses becomes important. Distractors should all be plausible, at least to a student who doesn't have the requisite knowledge or skills to complete the item.

Consider the examples shown here in Fig. 3. Both items reflect the science standard that calls for an understanding of atomic structure, but Fig. 3A, asks a student to provide an answer about an electronic configuration that can be readily memorized without much understanding of actual atomic structure, while the item shown in Fig. 3B also utilizes the construct of electron configuration but is more likely to show deep understanding of the concept. For teacher-designed assessments, this change in emphasis provides important additional information about student knowledge relative to more traditional test questions.

Figure 3. Examples of Items

What is the electron configuration of K^+?		Which electron configuration can only be for an atom in an excited state?	
(A)	$1s^22s^22p^6$	**(A)**	$1s^22s^22p^6$
(B)	$1s^22s^22p^63s^23p^6$	**(B)**	$1s^22s^22p^63s^23p^6$
(C)	$1s^22s^22p^63s^23p^64s^1$	**(C)**	$1s^22s^22p^63s^23p^54s^1$
(D)	$1s^22s^22p^63s^23p^64s^2$	**(D)**	$1s^22s^22p^63s^23p^64s^1$

Conclusion

Teaching and assessment go hand in hand for any instructor. While *No Child Left Behind* may have put new emphasis on certain forms of assessment such as high-stakes, large-scale testing, the fact remains that rich assessment informs good teaching. Chemistry teachers have a wide variety of resources available to them, some of which have been described here. As standards-based curricula become more important, assessments that measure learning within the guidelines established by the standards will also be key tools for teachers.

Recommended Readings

McMahon, M; Simmons, P; Sommers, R. *Assessment in Science: Practical Experiences and Education Research;* NSTA Press: Arlington, VA, 2006; 236 pp. For teachers who want the latest research about assessment techniques that work well, *Assessment in Science* is a good book. This collection includes reports by authors who are practicing K–12 classroom teachers, as well as university-based educators and researchers. Working in teams, they tried out and evaluated different assessment approaches in classrooms leading to sound research that nonetheless isn't hard to grasp. Several areas of assessment are covered under the categories of "classroom testing stories, standards-based assessment techniques, teaching-testing dilemmas, portfolio struggles and triumphs, and knowledge of the research on assessment."

W. James Popham, *Test Better, Teach Better;* Association for Supervision and Curriculum Development: Alexandria, VA, 2003; 175 pp. With the push for high-stakes testing associated with NCLB, it is tempting to dissociate concepts of assessment and teaching. This book provides evidence for how good testing can help with better instruction as well. In addition to more information about concepts like validity and reliability, the author of this book also provides rules for teachers to construct their own tests in ways that will provide valuable advice for teachers in their classrooms and for them to understand the results of their students on large-scale assessments as well.

Recommended Web Sites

ACS Division of Chemical Education, Examinations Institute, http://www.chem.iastate.edu/chemexams/ (accessed June 2008). This Web site contains valuable information about all of the ACS exams at all levels of chemical education. There are three exams available at the high school level: High School Chemistry Exam, Advanced High School Chemistry Exam, and the ChemCom Curriculum Exam, which has since 2006, been a noncalculator exam. There is also valuable information about national norms for each exam.

College Board AP Central, http://apcentral.collegeboard.com/apc/Controller.jpf (accessed June 20, 2008). This web site contains valuable information about all Advanced Placement Exams. Basic information about the AP Chemistry exam is available to all who connect to this Web site. AP Chemistry teachers can register on this Web site and obtain even more extensive information about the AP Chemistry exam, including past exam free-response questions and answers.

SAT Chemistry Subject Test, http://www.collegeboard.com/student/testing/sat/lc_two/chem/chem.html?chem (accessed June 20, 2008). This Web site describes information about the Chemistry Subject Test. On this site, you will find information about the skills that will be tested on this exam and recommendations for test preparation.

References

IBWeb 2007. http://ibchem.com/ (accessed Aug. 2007).

Nurrenbern, S.; Pickering, M. Concept Learning versus Problem Solving: Is There a Difference? *J. Chem. Educ.* 1987, *64*, 508–510.

Pellegrino, J. W., Chudowsky, N.; Glaser, R. Knowing What Students Know: The Science and Design of Educational Assessment. National Academies Press: Washington, DC, 2003. Popham, W. J. Test Better, Teach Better; Association for Supervision and Curriculum Development: Alexandria, VA, 2003; 175 pp.

Phelps, A. J.; LaPorte, M. M.; Mahood, A. Portfolio Assessment in High School Chemistry: One Teacher's Guidelines, *J. Chem. Educ.* 1997, *74*, 528–531.

Chapter 12

AP Chemistry: Course and Exam Review

by Jim Spencer and John Hnatow

Jim Spencer has a B. S. in chemistry from Marshall University and a Ph.D. in physical chemistry from Iowa State University. He is currently Emeritus Professor of Chemistry at Franklin & Marshall College. He is the 2005 recipient of the Pimentel Award in Chemical Education. He is currently chair of the Advanced Placement (AP) Chemistry Test Development Committee and co-chaired the AP Chemistry Course and Exam Review Commission for AP chemistry. Contact e-mail: jim.spencer@fandm.edu

John Hnatow received a B.S. in chemistry education from Kutztown University and a master's degree in environmental chemistry from East Stroudsburg University. John has taught chemistry for 36 years at Emmaus High School in Emmaus, PA, where he is also chairperson of the science department. John is an experienced Advanced Placement (AP) Chemistry consultant, AP and Pre-AP Workshop leader, AP Test Reader and Table Leader, and has served on the Chemistry Development Committee for four years. He co-chaired the AP Chemistry Course and Exam Review Commission for AP chemistry. Contact e-mail: jhnatow5@verizon.net

Introduction

In 2002, the National Research Council Center for Education, Division of Behavioral and Social Sciences and Education published a report: *Learning and Understanding: Improving Advanced Study of Mathematics and Science in U.S. High Schools* (National Research Council, 2002). The two-year study was based on current research upon learning and program design. The Advanced Placement, AP, and International Baccalaureate, IB, programs were evaluated, and recommendations were made on how these and other advanced study programs could become more effective and more accessible to students.

Among the principal findings of the NRC Chemistry Panel:

*   The AP and IB final examinations are formulaic and predictable in their approaches and question types from year to year.
*   Thus, with sufficient practice on how to take such examinations and enough drill on major concepts that the examinations are likely to test; students can score well on them primarily by rote, without actually understanding the major concepts associated with the topics being tested.
*   The AP and IB Chemistry courses to date do not yet recognize the increasingly interdisciplinary nature of modern chemistry; its incorporation of important related fields,

such as materials science and biochemistry; and the opportunities presented by such fields to teach related chemical concepts in a contextual manner.

- The AP and IB examinations do not reflect recent developments in chemistry and in the teaching of chemistry at the college/university level.

Subsequently, the NRC Chemistry Panel made these specific recommendations:

Any high school course in chemistry that is labeled as advanced study, whether it is structured according to an established curriculum and assessment (such as AP or IB) or otherwise, should enable students to explore in greater depth the chemistry concepts and laboratory practices introduced in the first-year course and, where appropriate, to conduct some form of research or independent inquiry. Under the guidance of a qualified advanced study instructor, desirable features of such advanced study would include some combination of these characteristics:

- application of basic ideas to complex materials, systems, and phenomena
- use of modern instrumentation, methods, and informational resources
- integration of concepts within and between subject areas, including extensions to other disciplines
- use of appropriate mathematical and technological methods
- extended use of inquiry-based experimentation
- development of critical thinking skills and conceptual understanding
- use of appropriate assessment tools of student performance that reflect current best practices
- promotion of communication skills and teamwork.

To be effective, advanced courses in chemistry must reflect recommendations in the areas of content, pedagogy, and assessment as described in the *National Science Education Standards* (NSES) (National Research Council, 1996).

Overarching Goals

The College Board immediately reacted to these findings and recommendations by undertaking a study of how to promote learning with understanding. Representatives of the College Board began consultation with the National Science Foundation (NSF) on possible changes that could be made to improve the learning environment in the sciences. Although AP courses and exams are successful imitations of typical introductory courses at colleges, the College Board has decided that it is insufficient for AP courses simply to reflect existing college courses if those courses do not represent instructional practices which promote deep learning and conceptual understanding. The College Board was urged to take AP one step further to emulate only those college courses that reflect the NSES standards. Up-to-date information is available on the College Board's AP Central Web site at http://apcentral.collegeboard.com/apc/Controller.jpf (last accessed March 21, 2008).

The AP program is subject to certain existing constraints:

- As a credit-by-examination program, AP must prepare students to succeed in sequent courses in each subject area.
- AP exam scores must be comparable to introductory course grades for predicting student performance in sequent courses.
- Any proposed redesign must maintain AP's ability to meet these requirements.

Within these existing constraints, the College Board has committed to the following guiding principles for the development of a revised AP Chemistry Course and Exam:

- Learners have different strategies, approaches, patterns of abilities, and learning styles that are a function of the interaction between their heredity and their prior experiences.
- Learners' motivation to learn and sense of self affect what is learned, how much is learned, and how much effort will be put into the learning process.
- The practices and activities in which people engage while learning shape what is learned.
- Learning is enhanced through socially supported interactions.
- Learning with understanding is facilitated when new and existing knowledge is structured around the major concepts and principles of the discipline.
- Learners use what they already know to construct new understandings.
- Learning is facilitated through the use of metacognitive strategies that identify, monitor, and regulate cognitive processes.

The National Science Foundation and the College Board undertook an ambitious effort with the sole purpose of aligning the AP Chemistry curriculum and the AP Chemistry Exam with introductory college courses that research has identified as those that best facilitate deep learning. The partnership between NSF and the College Board in the course and exam review of Advanced Placement science courses is based upon elements known as *overarching goals.* The overarching goals for a new paradigm in AP program development are to

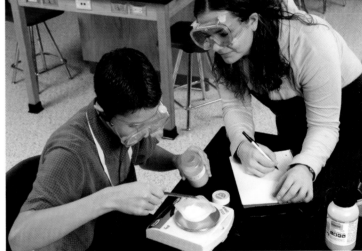

Mike Ciesielski

- increase depth of understanding of essential concepts while also developing capacity to use critical skills within the discipline by limiting breadth of content covered and drawing upon current research and theory on learning, instruction, and assessment to guide the design and implementation process
- infuse the AP science program courses with scientific topics drawn from cutting-edge research and emerging issues
- create science-learning programs accessible to students from a broad range of backgrounds, and
- prepare students for success in subsequent college-level courses within the STEM disciplines and stimulate them to consider careers in science, technology, engineering, and mathematics.

By meeting these goals, the objective of developing the habits of mind that support lifelong learning is addressed. Therefore, with NSF and the College Board's encouragement, an AP Chemistry Course and Exam Review Commission was formed. The Commission consisted of a diverse group of university researchers, chemical educators, and high school teachers. Some of the Commission members are experienced AP Chemistry Readers, Table Leaders, and AP Development Committee members. A listing of Commission members can be found at the end of this report.

The Commission convened four times during 2006–2007, worked between meetings, and applied the National Research Council's recommendations to the production of a new AP Chemistry curriculum, new exam specifications, and directed professional development opportunities. During the first half of 2008, several review panels were convened to review and refine the work of the Commission. These review panels consisted of university faculty, AP Chemistry high school teachers, and learning scientists. These reviews resulted in some of the following changes: terminology, reorganization, and the condensing and/or expanding of content topics and concepts.

How Students Learn

Changes to the AP Chemistry program will reflect the latest research on how students learn. The course and exam review will emphasize depth of understanding so that students will be better equipped to navigate complex content and to transfer their knowledge during assessments. A "less is more" principle, or the notion that it is better to uncover material than to cover it, will guide the selection of content. The major issues being addressed are breadth vs. depth and how to facilitate the development of advanced learners who choose to pursue additional study in chemistry. At present, there are more topics in the AP and introductory college courses than students can be expected to learn, and no topic receives adequate attention and depth. The solution involves altering the course material and process goals by cutting exercise solving and simple solving of numerical problems from the course. Students both in the lab and lecture should be required to interpret and analyze data (Lloyd and Spencer, 1994). On the basis of the original work of the Commission and the recommendations of the Review Panels, curriculum and assessment models were created which effectively emphasize problem solving, synthesis, and evaluation. The AP course will, to a higher degree, promote conceptual understanding and make connections within the discipline. Accordingly, the newly formed AP Chemistry Curriculum Development and Assessment Committee (CDAC) supervises any subsequent additions and/or revisions to these models.

The long-term goal is to increase scientific literacy and to encourage more students, especially those from groups traditionally underrepresented in the sciences, to pursue advanced-level study in high school and college and, eventually, to pursue science-related careers. All change is intended to help students move seamlessly from high school to college.

The new curriculum model will advocate college courses that research identifies as those with deep learning and conceptual understanding. The AP program has a concern that the amount of content on AP exams is putting inappropriate pressure on teachers to sacrifice depth of study to breadth of coverage. The AP program also has a concern that inquiry-based science learning is not being fostered. The inquiry approach to lessons is less authoritative and formal, more learner friendly, and closer to how scientists operate in the "real" world (American Chemical Society, 2003).

The College Board supports an AP Course and Exam Review that includes curriculum, instruction, and assessment. According to the NRC Report (National Research Council, 2002), a curriculum for high school advanced study

- structures concepts, factual information, and procedures that constitute the knowledge base of the discipline around organizing principles (unifying themes) of the subject area.
- links new knowledge to what is already known by presenting concepts in a logically sequenced order that builds on prior learning;
- focuses on depth of understanding rather than breadth of content coverage by providing students multiple opportunities to practice and demonstrate what they learn in a variety of contexts;
- includes structured learning activities that allow students to experience problem solving and inquiry in situations that are drawn from personal experience and real-world applications;
- develops students' abilities to make meaningful applications and generalizations to new problems and contexts; and
- incorporates language, procedures, and models of inquiry and truth verification that are consistent with the accepted practices of experts in the domain.

Accordingly, science courses

- should maintain students' focus on the central organizing themes, the underlying concepts of the discipline, and the unifying themes that engender further understanding;
- are based on careful consideration of what students already know;
- should focus on detecting, making visible, and addressing students' often fragile, underdeveloped understandings and misconceptions;
- should engage students in worthwhile tasks that provide access to powerful ideas and practices;
- should structure learning environments so students can work collaboratively to gain experience using ways of thinking and speaking used by experts in the subject area; and
- should orchestrate classroom discourse so students can make conjectures, present solutions, and argue about the validity of claims.

In addition, effective assessments are

- based on a model of cognition and learning that is derived from the best available understanding of how students represent knowledge and develop competence in a domain;
- designed in accordance with accepted practices to ensure reliability, validity, and fairness;
- aligned with the curriculum and instruction that the assessment is intended to measure;
- designed to include important content and process dimensions of performance in a subject area and to elicit the full range of cognition;
- multifaceted and continuous when used to assist learning by providing multiple opportunities for students to practice their skills and receive feedback about their performance; and
- designed to assess understanding that is both qualitative and quantitative in nature.

David Armer, USAFA

College Curriculum Study

In 2006, a content expert panel designed an instrument that identified instructional strategies that facilitate deep learning and conceptual understanding in existing college chemistry courses. The panel's selection of these strategies involved an invitation to 851 colleges and universities and a final review of faculty at 400 of these institutions. Experts were asked to nominate faculty who have been recognized for exemplary work teaching the general chemistry courses.

Instructors rated their own courses against the criteria of the instrument. The instrument was administered at colleges and universities where syllabi, assignments at all grade ranges, and exam items were collected.

Then criteria-based ratings were compiled, and nominees from prominent organizations and stakeholders analyzed the collected data and materials. These organizations and stakeholders included NSF, Science Education for New Civic Engagements and Responsibilities, Association for the Advancement of Science, National Association of Biology Teachers, American Institute of Biological Sciences, American Chemical Society, National Association of Environmental Education, AAAP, the AP Development Committee, and the College Board's Science Academic Advisory Committee.

As a result, the College Curriculum Study defined instructional quality in terms of three areas: (1) content taught, (2) habits of mind produced, and (3) teaching strategies and methods. A review panel identified twenty courses as excellent examples of having this type of instructional quality. Fifteen of the twenty courses were then advanced as exemplary college courses based on additional criteria, which included inquiry type learning, active involvement of students, integration of technology, etc. Courses and materials were compared to the instrument to ensure that they facilitated and promoted deep learning and understanding. Following this procedure and on the basis of a report submitted by the Center for Educational Policy Research (Conley, 2006), four courses were identified as exemplary. The College Curriculum Study also produced *performance expectations* (PEs), statements that reflected both conceptual understanding and reasoning and inquiry within each discipline.

The AP Course and Exam Review Commission members analyzed the discipline domain to gather, organize, and prioritize information that had implications for assessment. This was done in order to arrive at a better understanding of how the unifying concepts were expressed within the discipline and to find and describe the best ways to incorporate reasoning and inquiry. Completion of these tasks has informed the production of the following classroom supports: *AP Chemistry Course Guide*, embedded classroom assessments, new exam specifications, and professional development opportunities.

The Commission focused on what would be considered as evidence of student mastery of concepts. How will the instructor know what constitutes mastery and understanding? For this evidence, the college curriculum study documents can provide examples. This evidence-based part of the redesign of AP chemistry helped to establish course content (Mislevy and Riconscente, 2005). The Commission recommended what information is critical for inclusion in an AP course. Curriculum and assessment, as well as professional development working groups, will work cooperatively to decide how to measure what is taught in an AP context with its particular constraints, as well as how to support AP teacher instructional practices.

Unifying Themes

A domain (subject area) analysis model was developed as a conceptual tool that organized the knowledge, skills, and abilities that should be covered in the AP course. (Incidentally, this was the same organizing structure used across the sciences to highlight similar skills and common content among seemingly unrelated disciplines.) The advantage of the model is that it emphasized connections and helped the commission to move away from viewing the course as a long list of topics.

A preliminary model was shown to the principal investigators from the College Board, who suggested changes, and a draft model was discussed at a meeting of science co-chairs who made changes and approved the approach. The domain analysis model provides a course goal of developing integrated knowledge. It includes disciplinary knowledge organized around unifying themes, the ability to apply concepts, the understanding of connections between facts and concepts, the integration of concepts across the sciences, the ability to use inquiry to gain new knowledge, and the ability to reason scientifically.

With the goal of developing a new curriculum, a domain analysis ensued. This consisted of a detailed description of the structure of knowledge in the discipline. In order to accomplish this, the Backward Design process for curricular design was adopted (Wiggins and McTighe, 1998). Backward Design provides a method of systematically uncovering content by asking essential questions that focus on the big picture, goals for student learning, and pedagogical practice. This methodology begins with an identification of desired results followed by evidence that shows mastery of the concept. First, the linchpin ideas, that is, principles that transcend the discipline, were identified. The College Board recognized the ideas essential to understanding as unifying themes which served as hubs for the Backward Design. A unifying theme is one of a small number (five or six) of the most important organizing principles of the content within the discipline. Identified as "Level one," the unifying themes are conceptual anchors that promote coherence of knowledge and transference. A unifying theme is not to be considered as another fact or an abstract idea; it is to be a pedagogical tool that unifies knowledge and sharpens understanding. The goal is to reduce breadth and increase depth of understanding of the unifying themes. Several considerations related to them were discussed: (1) the overarching ideas that must be present if this course were the student's only opportunity for college-level study in chemistry, (2) the unifying ideas that must be present if a student will major in chemistry, (3) how such a foundation will serve the needs of subsequent courses, and (4) how students would demonstrate an appropriate depth of understanding of these unifying themes.

A sublevel, "Level two," contains a secondary concept, smaller in scope and related to the unifying theme, which will persist beyond the AP course. Level two consists of enduring concepts, what the student should take away from the course. These Level two concepts are essential knowledge that results from prioritizing and identifying those skills and concepts that have connective and transfer power. These are concepts that must be mastered, so students can perform complex tasks that demonstrate full understanding of the unifying themes at Level one. In the original work of the Commission, 30–40 Level two concepts were identified. The Level two concepts were too broad for course developers and test designers to identify what is and what is not in the course.

Therefore, a third level was required. In 2008, the Level two concepts were reviewed and refined by the Review Panels.

Level three consisted of concepts even smaller in scope and follows from Level two concepts. This level was designed to provide classroom and laboratory experiences that gave depth and understanding to the enduring concepts.

Level three was based on the essential knowledge required to support understanding of Level two. In order for a concept to be included at Level three, it must be integrated with the unifying concepts and scientific inquiry and reasoning.

The Commission and several review panels have identified, reviewed, and refined six unifying themes in the subject area. The unifying themes serve to guide the course and exam review. A draft of these unifying themes is listed in Fig. 1.

A graphic organizer tool called Inspiration provided commission members a visual representation for the mapping of the domain, and a common language to use among domains. Inspiration assisted the process of the domain analysis by emphasizing the connections of disciplinary knowledge to unifying concepts and inquiry throughout the process. A draft illustration of the use of Inspiration to map Unifying Theme 3 can be found in Fig. 2. This map shows how the unifying themes link to the enduring concepts, Level two, and how the enduring concepts may be strengthened and deepened by Level three.

A Level two concept labeled 3.A in Fig. 2 states that "Chemical changes are represented by a balanced chemical reaction that identifies the ratios with which reactants react and products form." The Commission decided by consensus to remove the rote memorization of solubility rules for reaction prediction exercises and documented this decision process using a tool called the "Reporter." (The *Reporter* is a tool that was developed by the College Board to be used as a product of the domain analysis).

The Reporter addressed and documented the reasoning needed to answer the questions found in the Commission's charge. This documentation was provided to subsequent review panels and the Curriculum Development and Assessment Committee, so that the reports could be used to develop the exam and professional development components of the course. This documentation also included decisions made with a consensus process

Figure 1. *Draft* of Unifying Themes to Guide the AP Chemistry Course and Exam Review

Unifying Theme 1:	The chemical elements are fundamental building materials of matter, and all matter can be understood in terms of arrangements of atoms that always retain their identity.
Unifying Theme 2:	Chemical and physical properties of materials can be explained by the structure and the arrangement of atoms, ions, or molecules and the forces between them.
Unifying Theme 3:	Changes in matter involve the rearrangement and/or reorganization of atoms and/or the transfer of electrons.
Unifying Theme 4:	Rates of chemical reactions are determined by details of the molecular collisions.
Unifying Theme 5:	The laws of thermodynamics describe the essential role of energy and explain and predict the direction of changes in matter.
Unifying Theme 6:	Any bond or intermolecular attraction that can be formed can be broken. These two processes are in a dynamic competition, sensitive to initial conditions and external perturbations.

to omit certain areas of content from the course. In Level three concept 3.A.3, another documented decision in the Reporter referred to the determination of empirical and molecular formulae. The Reporter recommendation served to exclude the traditional combustion analysis problems, which are rarely done today, but to include specific laboratory activities designed to allow students to determine simple formulas from the results of their experimentation. There are similar third-level concepts for 3.B and 3.C.

Figure 2. Draft of Unifying Theme Map for "Changes in matter involve the rearrangement and/or reorganization of atoms and/or the transfer of electrons."

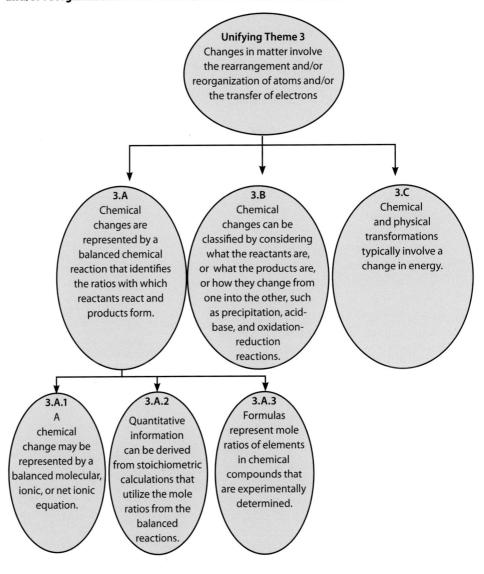

As a result of the work of the Commission, Review Panels, and College Readiness Products and Services Group, the AP Chemistry Course has been revised and redefined. The new AP Chemistry course is not intended to be a first formal opportunity to study chemistry. It bears reminding that learners construct their understanding of the world informally through their own experiences. Often, these explanations conflict with essential ideas within the discipline. Prior knowledge, skills, and abilities for taking an AP course have been identified, reviewed, and redefined. A draft of the topics not needed for understanding at the introductory level was also compiled, reviewed, and revised.

Also part of the AP Review were discussions regarding preconceptions and misconceptions that students are likely to bring to this course.

The first phase of the AP Chemistry Course and Exam Review was completed in 2007. Subsequent review panels were formed and convened to review and refine the initial work of the Commission. Once the review panels completed their work, the Curriculum Development and Assessment Committee supervises any additions and/or changes to the curriculum framework. This Committee also serves as the College Board's spokespersons, sharing this work with professional organizations, higher education faculty, and AP high school teachers alike. The new AP Chemistry course and exam will promote professional development opportunities, as well as provide updated classroom resources.

AP Chemistry Redesign Commission

John Hnatow, Emmaus High School (Co-Chair)
Jim Spencer, Franklin & Marshall College (Co-Chair)
Sean Decatur, Mt. Holyoke College
John Gelder, Oklahoma State University
Reen Gibb, Brookline High School
Carlos Gutierrez, California State University, Los Angeles
Annis Hapkiewicz, Okemos High School
Thomas Holme, Iowa State University
Jennifer Kennison, North Country Union High School
Dana Krejcarek, Kohler High School
Luisa Marcos, Union Hill High School
Angelica Stacy, University of California, Berkeley
Becki Williams, Richland College

College Board – Advanced Placement Program

Marcia Wilbur, Executive Director, Curriculum and Content Development
John Eggebrecht, Director, AP Science Redesign
Tanya Sharpe, Associate Director, Curriculum and Content Development

Acknowledgments
National Science Foundation under Grant ESI 0525575
James Pellegrino, PI, University of Illinois
Jeanne Pemberton, Co-PI, University of Arizona
Mark Reckase, Co-PI, Michigan State University

References
American Chemical Society. *Chemistry in the National Science Education Standards;* American Chemical Society: Washington, DC; 2003.

Conley, D. T. Advanced Placement Best Practices Study—Year 1. University of Oregon: Eugene, OR, 2006.

Lloyd, B.; Spencer, J. *New Directions for General Chemistry*. Division of Chemical Education with funding from the National Science Education. American Chemical Society: Washington, DC, 1994.

Mislevy, R. J.; Riconscente, M. M. (2005). *Evidence-Centered Assessment Design: Layers, Structures, and Terminology;* PADI Technical Report 9; SRI International: Menlo Park, CA, 2005.

National Research Council (NRC). *Learning and Understanding: Improving Advanced Study of Mathematics and Science in U.S. High Schools*. Center for Education, Division of

Social Sciences and Education. National Academies Press: Washington, DC, 2002.

NRC. *National Science Education Standards*. National Academies Press: Washington, DC, 1996.

Wiggins, G.; McTighe, J. *Understanding by Design;* Association for Supervision and Curriculum Development: Alexandria, VA, 1998.

Chapter 13

Embracing Diverse and English Language Learners in Chemistry

by Doris Kimbrough and Susan Cooper

Doris Kimbrough *graduated from Cornell University with a Ph.D. in organic chemistry. She joined the chemistry faculty at University of Colorado, Denver, in 1986, and is currently principal investigator for the NSF-funded Rocky Mountain-Middle School Math and Science Partnership, which provides professional learning for middle-level math and science teachers. Contact e-mail: doris.kimbrough@cudenver.edu*

Susan Cooper *received her Ed.D. in curriculum and instruction from the University of Central Florida in 2004. She taught chemistry and physics and served as science department head for 27 years at LaBelle High School in LaBelle, FL. In August 2007, she began a new career as assistant professor of education at Florida Gulf Coast University. Contact e-mail: sjcooper@fgcu.edu*

Introduction

The *National Science Education Standards* (1996) stipulate that science education policies must be equitable for *all* students. Chemistry teachers should be preparing their students for a multicultural world by celebrating diversity in an inclusive classroom environment, but this is not as easy as it sounds. Four decades after the initiation of equal rights and protection legislation, the number of baccalaureate degrees awarded to women in science and engineering has finally reached parity with that awarded to men; however, the number of female students proceeding on to graduate school and getting post-baccalaureate degrees, while increasing, still lags behind the number of men (NSF, 2004). Moreover, the percentage of science and engineering degrees awarded to minority students is less than that for the general population, and this difference is greatest in the physical sciences, the category that includes chemistry (NSF, 2004). This divergence increases as one looks beyond the bachelor's degree to the percentage continuing on to graduate school. The reasons for these differences are complex, debatable, often contentious, and not easily solved by a single classroom teacher struggling to help her students understand stoichiometry and chemical bonds.

JupiterImages

Equity of access to challenging curriculum in mathematics and science can often serve as a kind of negative feedback loop. Students who are members of racial or ethnic groups that are traditionally underrepresented in the sciences are often not encouraged to pursue those sciences, which results in those groups continuing to be disproportionately underrepresented. There are many advantages to including all students in instruction, beyond the fact that all students need a basic knowledge of chemistry to make personal and community decisions in our technological society. These advantages include the accommodations that help all students learn chemistry and the unique perspectives that diverse individuals can provide in classroom discussions (Miner et al., 2001). As teachers, we must ensure that students who choose not to pursue science are doing so because they lack interest and enthusiasm for those disciplines, ***not because they are discouraged***. Often, lack of encouragement can be as detrimental as outright discouragement. We must actively promote gender equity, and we must actively promote cultural equity in the examples and materials we use in our classrooms. According to the *National Science Education Standards* (National Research Council, 1996), how we teach and our relationships with our students greatly influence what our students learn.

This chapter can only scratch the surface of what are deep and complicated issues around increasing population diversity, access, retention, the achievement gap, and the pressures on school systems to accomplish more and more with ever diminishing support. This chapter offers an introduction to the multitude of strategies that have proven effective in the teaching and encouragement of diverse populations. Readers are encouraged to go beyond this chapter to explore the vast literature that is available, more directly focused on these challenges, and far more thorough than this simple treatment. Chapter 15 in this volume is an excellent resource for more information about how all students learn.

Teaching Chemistry to English Language Learners

The 2000 U.S. Census reported that over 18% of our population speaks a language other than English in their home, and in some states, the percentage is far higher. School age children comprise a large number of these English Language Learners, putting a strain on many state and local school districts as they struggle to meet the needs of this ever-growing population. In this chapter, we choose to use the term English Language Learner (ELL), as it is a more accurate designation than the expression, English as a Second Language (ESL), as many of these students already speak more than one language and are acquiring English as their third, fourth, or even fifth language. ELL students hail from around the world, and educators can no longer rely solely on bilingual Spanish or Asian language speakers to accommodate ELL students. Achievement efforts of ELL students are often additionally hampered by limited literacy in their native language, which makes achieving fluency and literacy in an acquired language far more difficult.

Certainly, this chapter will address working with ELL students from a science/chemistry perspective; however, it should be stressed that what constitutes good instruction for ELL students is really good instruction for *all* students, even native English speakers. All students can benefit from strategies such as concept maps, breaking complex concepts into smaller, more comprehensible ideas, or providing more than a single explanation or illustration of a concept. Structuring lessons to support ELL students also supports students with mild learning disabilities, alternative learning styles, different content backgrounds, or different cognitive levels. Science and mathematics in general, and chemistry, in particular, offer a unique opportunity to further the knowledge and understanding of ELL students because of our heavy reliance upon symbolic rather than verbal representations. This, coupled with the technical vocabulary that accompanies the study of chemistry—new for *all* students—provides ELL students a more level playing field for learning alongside their native English speaking peers. Unfortunately, in many school settings the opportunity to collaborate for teachers of science and ELL specialists does not exist, so it is not surprising that most chemistry teachers are a bit mystified about how to teach chemistry to a student with limited or no English skills. We offer some suggestions here, but we also strongly encourage teachers with access to ELL specialists within or outside of their school districts to avail themselves of this resource through professional development activities. Most of the instructional strategies used to teach ELL students can be adapted for use in all content areas, including chemistry. In addition, beginning chemistry teachers should seek out other science, mathematics, and technology teachers with experience teaching science to ELL students. Coordinating your colleagues' experience with advice from ELL specialists will empower you to meet all of your students' needs.

Engaging All Students

Structuring a lesson or a curriculum to make it more accessible to ELL students is often referred to as "sheltering" or "scaffolding" (Echevarria and Graves, 2007). Sheltered instruction in this context no longer suggests that students are protected from the higher-level content their native English-speaking peers' experience, just that the content is presented in a way where the learner feels safe and able to connect the content to previous conceptual understanding. Teachers must explicitly demonstrate their expectations and frequently, informally assess students so that the scaffolding can be adjusted to help the students become successful learners (Galguera, 2003).

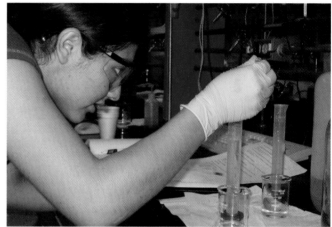

Melanie Haas, CSU MESA Director

After a professional learning session regarding proven strategies for teaching ELL students, a colleague made the comment that, "This isn't about sheltered instruction; it's just *good* instruction!" Indeed, many of the strategies that help the struggling ELL student will also work for native English speakers, particularly those that are nonverbal learners, struggle with learning disabilities, or have poor backgrounds in science or mathematics. In spite of the fact that it is the preferred mode of delivery for many of our colleagues, lecturing over content is not the best way to connect with many of our students, particularly students with limited English proficiency. Varying how students can access content will help to engage all of your students, regardless of language or other barriers. Mixing in lab or demonstration activities, group discussions, problem-solving activities, and other nonlecture forms of presenting content will provide the best venue for all learners. The annotated list below contains suggested strategies that literature has shown to be effective in teaching ELL students; there is also literature that supports many of these strategies for students who are visual or kinesthetic learners and for students who have particular learning disabilities (e.g. sensory integration disorders, attention deficit disorders, or several of the autism spectrum disorders). It is by no means a complete list of all possible strategies, nor are we suggesting that every teacher incorporate every strategy. It is best to find strategies that fit one's personal style and student population. These approaches are supported by Teaching Standard B in the NSES: Teachers of science guide and facilitate learning (National Research Council, 1996).

- **Organize instruction.** Establish routines for bell work, lab activities, reading and note taking, and class participation. Explicitly model and communicate your expectations and practice the routines with your students.

- **Plan ahead.** Provide a syllabus and due dates for assignments, then stick to your plan. Use technology to post information on the Internet. Include class notes and extra information such as links to videos and useful Web sites.

- **Cues to verbal information.** If you have ever had the experience of conversing in a language in which you are not 100% fluent, you will know that it can be exhausting. The words fly by at an alarming rate and often seem to run together into meaningless gibberish. A good teacher trying to help her ELL students access her explanations will speak clearly, avoiding too much slang (or at least explaining it!), and accompany her words with additional cues. These cues can be motions that act out particular words or phrases: e.g., using your hand to make a talking motion if you want your students to engage in discussion; writing motion in the air when you want them to record information, or using index and middle finger to point to your eyes and then moving outward to indicate the act of making observations. Cues can also be actual props or pictures that support vocabulary. Hold up the beaker as you discuss it. Point to the 50-ml mark on the side as you instruct them to fill it with 50 ml of water. Show a picture or a model of whatever concept you are trying to explain. Face students as you speak. Stick to the topic. Repeat questions asked by other students to make sure everyone heard and understood the question. Minimize distractions, especially extraneous noise in the classroom.

- **Idioms and analogies.** Avoiding slang, as ELL specialists advise, is a particular challenge, particularly since slang and phrases that have pop-cultural pertinence will help you connect to teens and preteens and will make science and chemistry relevant to your students. Being aware of your slang use and calling attention to it will help your ELL students: *What do I mean when I say "in the ballpark"? What does 24/7 mean? If we are "not on the same wavelength", are we communicating or not communicating?* Similarly be aware that analogies that make sense to us as Americans may be nonsensical to those who are raised in different cultures. For example, using a ham sandwich analogy to illustrate limiting reagents will not make sense to a student raised on rice and noodle dishes. Analogies can be very helpful in helping students connect to conceptually difficult or abstract material, but research cautions instructors not to rely too heavily on analogies and to point out to students the limits of utility of an analogy (Orgill and Bodner, 2004). Using a staircase model to illustrate quantum levels for electrons may lead to the misconception that the energy separating different levels is constant.

- **Multiply your explanations.** By "multiply," we mean both the act of explaining a concept more than once and elaborating the explanation to include multiple modes. For example, an explanation of an ionic substance dissociating as it dissolves in water could include a verbal explanation of what is happening, representational descriptions of ions and molecules (Na^+ and Cl^- ions scattered among H_2O molecules), circular or spherical models of ions surrounded by the "Mickey Mouse" depictions of water molecules, and a reiteration of the entire explanation using each mode again. Try to reach all modalities (kinesthetic, visual, oral) of learners in your classroom.

- **Vocabulary.** Many "best teaching practices" proponents disparage the use of vocabulary lists as being part of the "science as disconnected facts" approach rather than the more inquiry-focused approaches favored today. However, as students move through their study of chemistry, the vocabulary gets more complex and demanding, particularly for the ELL student. Lists of vocabulary words have their place in instruction. Going through the list and having your students repeat the words after you may seem like a pointless exercise, but even native English speakers will pronounce *cation* "CAY-shun" and struggle with the pronunciation of *protactinium* when encountering it for the first time. Many teachers ranging from kindergarten upward have had success with "word walls," which have new vocabulary posted on walls or bulletin boards in the classroom, typically with pictures. Review the words often with your students and encourage them to use them in their conversations and writing. Chemistry teachers should encourage their ELL students to look for cognates, especially those words that have roots in European languages. For example, "aqueous" is related to the Spanish word "agua." However, we caution that many false cognates also exist, so teachers must help students recognize those as well. One source of confusion that students newly arrived from a "metric"-speaking country might experience relates to our use of two measuring systems, English and metric. Furthermore in many countries that have Romance language roots, the cold water faucets are marked *F* (i.e., *frio* or *froid*) and hot water is labeled *C* (i.e., *caliente* or *chaud*).

- **Group work.** Chemistry is a collaborative discipline, and all students can benefit from peer interaction as they struggle with chemical concepts. Teaching Standard E (NSES, 1996) supports developing communities of science learners who collaborate in scientific inquiry. It is best to group students in varying and creative patterns. ELL students that share a common language can at times be grouped together for mutual support. Other times you may want to group or pair struggling students (ELL or otherwise) with their more successful counterparts. Similarly, students within groups should rotate tasks so that, for example, the girls are not always the recorders and that the ELL students are compelled to record data or write sentences. Group work can also build confidence and allow time for understanding: posing a question and then allowing time for group discussion before asking for responses from individuals or groups will foster participation among students who tend to be shyer, whether for linguistic or cultural reasons.

- **Evidence for claims.** To a student struggling both with the subject of chemistry and with a new language, how knowledge is acquired can be very mysterious and sometimes almost magical. Constantly revisiting how or why we know what we know is a useful way to illustrate the process of science. This can range from the mundane and straightforward (*and how do we know that HCl is a strong acid? Because we memorized the six strong acids!*) to the more esoteric (*why does ice float on liquid water? Because it's less dense. And why is it less dense? Because the crystalline form has empty spaces. And why are there empty spaces? etc.*) Teachers should also be sensitive to the differences between what is considered "good authority" in a scientific context as compared to other aspects of society. A discussion of the terms "law" and "theory" in a scientific context versus a societal context is a useful way of approaching this difference.

- **Literacy strategies.** Pick up a magazine or newspaper in another language. If you are familiar with that language, reading large chunks of text is challenging, even if there are pictures or diagrams to guide understanding. The annotated readings and Web sites at the end of this chapter contain dozens of literacy strategies that teachers can use to help guide students through nonfiction reading (graphic organizers, word searches, concept mapping, etc.). Many of these strategies will assist both ELL and native English speakers alike. Both beginning and experienced teachers can benefit from regular conversations with their colleagues who specialize in literacy. Reading science text requires students to recognize how graphs, charts, photographs, and other visual cues are used to convey scientific information. Making the connections between what is in the book and how students can improve their science skills requires explicit modeling by the teacher. *ChemMatters* magazine provides reading strategies in the online teaching guide for each issue. Most high school textbooks also provide reading strategies with the ancillary materials for teachers. Teachers should always give students a purpose for reading, such as a discussion to pique student interest in the topic, a written anticipation guide to complete prior to reading, or simply asking students to make predictions about what will be in the reading. Chemistry teachers should be encouraged to explore literacy strategies with language arts specialists who can offer a variety of tactics to foster effective nonfiction reading.

- **Technology.** For many students, access to technology is highly motivating. Teaching Standard D (NSES, 1996) supports the use of technology in teaching. Many technology devices, such as laboratory probes and computer programs, require minimal reading or prior instruction while allowing students to learn about scientific processes. (See chapter 7 for additional discussion on the use of technology in the chemistry classroom.) Use of technology also allows for differentiated instruction, offering multiple forms of representation, so that students can be given assignments that meet their diverse needs. For example, teachers can develop alternative assessments and allow students to choose how they would like to demonstrate understanding.

- **Make learning personal.** Look for ways to make chemistry relevant and meaningful to students' lives. In addition to career exploration, encourage students to find scientific contributions made by students like them, such as Alice A. Ball, an African-American woman whose research was instrumental in developing a drug to treat Hansen's disease. (See the February 2007 issue of *ChemMatters* for an excellent article describing her work.) Mario José Molina Henríquez, a chemist at the University of California at Irvine, is a native of Mexico and won the 1995 Nobel Prize in chemistry for his work in linking chlorofluorocarbons (CFCs) to the destruction of the earth's ozone layer. The Chemical Heritage Foundation has an excellent online exhibit about women in chemistry, "Her Lab in Your Life." In the ACS publication *Teaching Chemistry to Students With Disabilities*, you can find a periodic table describing the disabilities of several scientists who discovered elements. Even learning a few words in students' languages can make a huge difference in the accepting culture of the classroom. Teachers should plan to celebrate the achievements of chemists from diverse backgrounds throughout the school year so that students will see that there are no barriers to their own achievement. Students could research these achievements and share their newfound knowledge with the class through Power Point presentations or posters. In the process, they will learn about the history and nature of science (Content Standard G, NSES, 1996).

JupiterImages

- **Talk to your students.** Ask them how you can help them learn chemistry, and don't be afraid of what they might tell you. Acting upon their advice (within reason!) will empower and motivate their learning.

Assessment of Student Learning

No Child Left Behind (NCLB) requires that all cohorts, including ELL students and those with disabilities, improve their performance on math, reading, and science tests. ELL students are required to take the math test the first year they are in the United States, but they may be excluded from the reading/language arts test the first year (Fact Sheet, 2007). Accommodation policies for the ELL subgroup vary from state to state, with some states providing bilingual dictionaries for students and others providing limited translation on math tests. Since science tests are required nationwide beginning in the 2007–2008 school year, the different state accommodation policies are in flux. Also, adequate yearly progress (AYP) requirements vary from state to state. Because each state has developed its own science standards, teachers must familiarize themselves with their states' requirements. (Chapter 11 in this volume offers more information about assessment resources for chemistry.)

Teaching Standard C in the NSES (National Research Council, 1996) requires that teachers engage in ongoing assessment to evaluate their teaching, as well as student learning. Although classroom assessment should align with curriculum standards, that does not mean that all assessments must be objective tests. In fact, higher-level cognitive skills such as planning investigations, analyzing and synthesizing information, and critical thinking are often difficult to measure on these tests. In addition, there is no opportunity for students to communicate their ideas and lab results on objective tests. Teaching Standard C also calls for teachers to guide students in self-assessment, which can promote metacognition while motivating students to achieve at a higher level. Chemistry teachers should have high expectations for all students, even though they are given different opportunities to demonstrate conceptual understanding through authentic tasks such as projects, presentations, and portfolios. On their way to comprehension, students need frequent formative feedback from their teachers. This feedback can take the form of interviews, observations, and checklists. However, performance assessment in chemistry should not be done to the exclusion of reading and writing assessment because students must learn to be successful at both types of assessment to demonstrate scientific understanding (Kamil and Bernhardt, 2004). Consider making at least some classroom assessments visually, linguistically, and culturally organized to resemble the high stakes tests required by NCLB to lessen the confusion for students and to ensure that the assessments more accurately reflect the scientific knowledge of students (Luykx et al., 2007).

Parental Involvement

The involvement of parents in supporting student learning is essential to the educational success of diverse learners. Chemistry teachers should explicitly define how all parents can help their children learn chemistry. For example, teachers can provide information regarding the frequency and type of homework assignments and how to study chemistry. If you use rubrics, explain them to parents. If possible, books in Spanish should be provided to parents. When students know that their parents are supportive of what they are learning in chemistry, they will perform better. Even parents who have a limited education can motivate their children by providing an environment free of distractions and coping skills to help when their children become discouraged. Homework should be purposeful, encouraging family involvement whenever possible. Authentic learning experiences such as those provided in *Chemistry in the Community* (*ChemCom*) offer opportunities for students to relate what they are learning in chemistry to their particular circumstances. (See chapter 8 for a detailed discussion of how *ChemCom* is being used in inner city schools.)

If you teach students with learning disabilities, including those with ADHD (Attention Deficit Hyperactivity Disorder), they will have an IEP (Individualized Education Plan) that is developed by the teachers and the parents. The IEP will delineate accommodations that will help the student learn all subjects. Before an IEP meeting with parents, you should outline strategies that will facilitate student understanding in chemistry. Teachers should communicate to parents their high expectations for all students, including girls, minorities, and disabled students. Parents of all students should be made to feel welcome at the school. This can be accomplished through activities such as parent nights and parent workshops, where information regarding what students are learning in chemistry is shared with parents. Parents also want frequent communication with teachers, even if that means calling in translators. Schools with high numbers of ELL students ought to have translators on staff who can be called upon for parent/teacher conferences, as well as community liaisons who can aid dialogue between the school and home. Bilingual teachers, paraprofessionals, or community members can translate letters, posters, and other written materials that have information about chemistry courses and the importance of learning chemistry. When parents feel welcome at the school, they can come to events and their children can translate when necessary.

Finally, the NSES Teaching Standard C (National Research Council, 1996) calls on teachers to use student data to communicate with parents regarding both student achievement and the *opportunities* they have to learn science. Although we routinely report student achievement to parents, parents of students traditionally underrepresented in the sciences may not realize that their students have the opportunity to take chemistry. Therefore, it is our task to ensure that parents understand that their children have the ability to succeed in chemistry. By sharing success stories of diverse students and clearly communicating expectations for chemistry courses with parents, teachers can gain support and reach more students.

Community Involvement

The community is another rich source of support for diverse learners that should be included in the school science program (Teaching Standard D, NSES, 1996). Chemistry teachers should find opportunities to use community examples that demonstrate the importance of chemistry in our everyday lives. Most students know that they need chemistry to work in the health care industry, but they may not realize that they also need knowledge of some basic chemistry to work as cosmetologists, mechanics, cooks, artists, and more. Business owners are potential employers who can emphasize the science requirements to work for their companies. Many rural communities with diverse learners have an agricultural focus. Agricultural communities, in particular, need high school graduates who have a rudimentary understanding of chemistry in order to make basic decisions regarding safety and efficacy when using chemicals on crops. Chemistry teachers and agriculture teachers can work together to promote student understanding. For many jobs that students might consider after high school, being a member of a diverse population may be an advantage. For example, knowledge of two or more languages helps health care workers in diverse communities relate to their patients.

Encouraging traditionally underrepresented students to pursue careers in scientific and technological fields can take a number of different forms in a classroom setting. Job markets are traditionally good in most of these fields, so students can see their education dollars efficiently converted into earning dollars. Many chemical careers relate to fields that have an altruistic component, such as medicine or the environment. Students interested in "making a difference" or "helping their fellow man" may be drawn to these careers, especially when examples are tied directly to those cultural communities where the student feels most at home. Even though many chemists are white males, diversity of race, gender, and ethnicity can be found among our ranks, and as teachers of diverse student populations, we should be particularly proactive at promoting chemists who are women, Black, Latino, Asian, Native American, and/or physically disabled.

Field trips, guest speakers, and mentors can help students see that chemistry is practiced by many different people, including women and minorities. Guest speakers might include college students majoring in chemical fields who can offer new perspectives. Service learning tied to what they are learning in chemistry is another avenue that chemistry students could explore in their communities. Students could set up recycling centers, organize after school science programs for elementary students, or perform water quality testing in their communities. Through programs such as this, students learn that chemistry is relevant to their lives (Content Standard F, NSES, 1996).

Field trips can also help to provide all students with the same background knowledge, including students newly arrived in the community. When studying energy use, for example, visiting a local museum to learn more about energy use in the community 50 or even 100 years ago gives all students the same setting to refer to for comparison to today's energy use. Field trips do not necessarily require that students travel long distances. Some excellent resources for learning about chemistry may be right in the community, such as water treatment plants, bodies of water where water quality can be studied, or health care facilities. While on field trips, make the learning personal by asking students how the activity impacts their lives (other than getting them out of the classroom!).

At the school and district levels, the *National Science Education Standards* (1996) includes program standards that describe conditions that must be in place for a comprehensive program that provides all students access to learning science. In particular, Program Standards D and E stipulate that all students have adequate, safe space to conduct scientific inquiry both inside and outside the school building. As teachers, we should advocate for our students to ensure that necessary accommodations are made to provide the opportunity for *all* students to learn chemistry.

Conclusion

In order to meet the goals of the *National Science Education Standards* (1996) regarding equitable science education policies for all students, chemistry teachers should be open-minded and flexible and act as advocates for their students (Cline and Necochea, 2006). Teachers can become reflective practitioners that affirm diversity. Most importantly, science teachers should have high, yet realistic, expectations for all of their students. Through these practices, we can begin to reach the goal of educating a scientifically literate society in which all students have the opportunity to learn science and contribute to our understanding of the world.

Recommended Readings

Keeley, P.; Eberle, F.; Farrin, L. *Uncovering Student Ideas in Science: 25 Formative Assessment Probes*. NSTA Press: Arlington, VA, 2005. The authors describe formative assessments, how to use the probes, and links to the *NSES* and *Benchmarks for Scientific Literacy*. Many of the probes provided are directly related to chemistry.

The Science Teacher (March 2007 and March 2008). Both of these issues are focused on teaching diverse students. They include many practical examples, as well as references.

Thier, M. *The New Science Literacy: Using Language Skills to Help Students Learn Science*; Heinemann: Portsmouth, NH, 2002. This book provides many examples and reproducible pages that combine science, language, and guided inquiry in order to facilitate students' growth as independent learners.

Recommended Web Sites

Chemical Heritage Foundation Home Page. http://www.chemheritage.org (accessed March 16, 2008). From the home page, you can find many excellent classroom resources related to the contributions of many diverse chemists, including the online exhibit "Her Lab in Your Life."

Fact Sheet: NCLB Provisions Ensure Flexibility and Accountability for Limited English Proficient Students. http://www.ed.gov/print/nclb/accountability/schools/factsheet-english. html (accessed April 10, 2008). Questions regarding the testing of limited English-proficient students in language arts and mathematics are answered at this Web site.

Helping English Language Learners in the Science Classroom. http://teachingtoday.glencoe. com/howtoarticles/english-language-learner-teaching-strategies-that-work (accessed March 16, 2008). This commercial Web site has many specific strategies for helping all students, not only English language learners (ELL) students, learn science.

Science and Technology Literacy. http://www.ncela.gwu.edu/resabout/literacy/3_content/ 3_science.htm. (accessed March 16, 2008). This Web site of the National Clearinghouse for English Language Acquisition lists many resources for helping ELL students learn science and technology.

SDAIE Strategies: A Glossary of Instructional Strategies. http://www.suhsd.k12.ca.us/suh/
---suhionline/SDAIE/glossary.html. (accessed March 16, 2008). *SDAIE* is an acronym
for *S*pecially *D*esigned *A*cademic *I*nstruction in *E*nglish, so this is a list of instructional
strategies to use with ELL and other students.

*Teaching Chemistry to Students With Disabilities: A Manual for High Schools, Colleges,
and Graduate Programs.* Available online at http://membership.acs.org/C/CWD/
TeachChem4.pdf (accessed April 10, 2008). This booklet, also available free of charge
from the American Chemical Society at 1-800-227-5558, includes excellent information
for teaching students with physical and learning disabilities. There is a chapter devoted to
lab accommodations, as well as an extensive resource list for more information regarding
specific disabilities.

Resources for Teaching ELL Students

Echevarria J.; Vogt, M.; Short, D. *Making Content Comprehensible for English Language
Learners: The SIOP Model*, 2nd Ed.; Pearson Publishing: Boston; 2004.
Echevarria J.; Graves, A. *Sheltered Content Instruction: Teaching English Language
Learners With Diverse Abilities*, 3rd ed.; Allyn & Bacon: Boston, 2007.

References

Chemistry in the Community. http://www.whfreeman.com/chemcom (accessed April 10, 2008).
ChemMatters. http://www.acs.org/chemmatters (accessed April 10, 2008).
Cline, Z.; Necochea, J. Teacher Dispositions for Effective Education in the Borderlands.
The Educational Forum 2006, *70*, 255–267.
Echevarria J.; Graves, A. *Sheltered Content Instruction: Teaching English Language
Learners with Diverse Abilities,* 3rd ed.; Allyn & Bacon: Boston, 2007.
Galguera, T. Scaffolding for English Learners: What's a Science Teacher to Do? *FOSS
Newsletter* #21 (Spring 2003). http://www.lhs.berkeley.edu/foss/newsletters/archive/
FOSS21.Scaffolding.html (accessed April 10, 2008).
Kamil, M. L.; Bernhardt. E. B. The Science of Reading and the Reading of Science:
Successes, Failures, and Promises in the Search for Prerequisite Reading Skills for
Science. In *Crossing Borders in Literacy and Science Instruction: Perspectives on Theory
and Practice;* Saul, E. W., Ed.; International Reading Association: Newark, DE, 2004;
pp 123–139.
Luykx, A., Lee., O., Mahotiere, M., Lester, B., Hart, J. and Deaktor, R. Cultural and Home
Language Influences on Children's Responses to Science Assessments. *Teachers College
Record* 2007, *109*, http://www.tcrecord.org (accessed March 31, 2007).
Miner, D.; Nieman, R.; Swanson, A. B.; Woods, M., Eds. *Teaching Chemistry to Students
With Disabilities: A Manual for High Schools, Colleges, and Graduate Programs,* 4th ed.;
American Chemical Society; Washington, DC; 2001.
National Research Council. *National Science Education Standards.* National Academies
Press: Washington, DC, 1996.
National Science Foundation. *Women, Minorities, and Persons with Disabilities in Science
and Engineering.* Division of Science Resources Statistics, National Science Foundation:
Washington, DC, 2004.
Orgill, M.; Bodner, G. What Research Tells Us About Using Analogies to Teach Chemistry.
Chem. Educ. Res. Pract. 2004, *5*, 15–32.

Chapter 14

Prior Knowledge of Chemistry Students: Chemistry K-8

by Dorothy L. Gabel and Karen J. Stucky

Dorothy L. Gabel *is Professor Emerita of Science Education at Indiana University. She earned her Ph.D. at Purdue University under the guidance of J. Dudley Herron. Her* Saturday Science *program and work on the* Indiana Core 40 Assessment *have impacted thousands of teachers and elementary school children. She has been honored with the Robert Carleton Award for Leadership in Science Education from the National Science Teachers Association and the Award for Achievement in Research for the Teaching and Learning of Chemistry from the American Chemical Society. Contact e-mail: gabel@indiana.edu*

Karen J. Stucky *is a retired teacher of upper elementary science and math. After retiring from the classroom and being the elementary science coordinator for our local school system, Karen became actively involved in informal museum education and professional development in science for teachers. She serves as Education Director at the WonderLab Museum of Science, Health and Technology, Monroe County Community School Corporation, in Bloomington, IN. She designs field trips, laboratory lessons, and teacher workshops at the museum. After a second retirement later this year, she will remain a consultant for the museum and for Delta Education doing FOSS training for teachers. Contact e-mail: kstucky@indiana.edu*

Introduction

A guest editorial on the Editor's Page in the August 28, 2006, issue of *C&E News* entitled "Forget Chemistry" (Wolke, 2006) contained an important message about chemistry instruction for children. The author expressed the opinion that

> *"In our efforts to swell the ebbing stream of students choosing chemical careers, we may be trying too hard. We may put our faith in exposing children to chemistry at an early age in hopes that they will catch the "chemistry bug" before it (and they) are barely out of the larval stage.... But the usual children's enticements often turn out to be little more than magic shows that, frankly, generate little or no excitement about careers in chemistry. Let's face it: Neither colored liquids or fizzing baking soda are as fascinating to kids as the insides of a salamander or a black hole. A general captivation by Nature must be instilled before we can expect children to "specialize" in chemistry."*

Students' continued interest in majoring in chemistry at the college level is dependent on their success at the high school level and in their introductory college-level chemistry courses. Even if children "have been captivated by nature" at the elementary level, this is insufficient to encourage them to select chemistry as a career. For some students, success at the high school level is insufficient because of the way high school chemistry is taught or because at the elementary or middle school level, the content was beyond their comprehension.

For example, a young woman who recently graduated from a large state university and who earned the highest grade in the introductory chemistry course of about 500 students was selected as the secretary/assistant for a summer chemistry program for high school chemistry teachers. Asked why she waited to take the chemistry course until after she had graduated from the university with a major in psychology, she replied, "I didn't understand a thing in

high school chemistry, even though I got an A in the course." She decided to major in psychology after examining a variety of textbooks in the university bookstore. She said that she found the content in the psychology books interesting and could understand the subject matter by reading! This young woman eventually wanted to become a veterinarian. After she graduated from college, only then did she feel confident that she would be able to pass the required chemistry and other science courses to fulfill her plans for the future.

The above account is not unusual. An excellent third-grade elementary teacher who is working part-time conducting science workshops for a publisher of one of the best science programs in the United States, was also turned-off to chemistry in college. She earned an "A" in an introductory chemistry course during her first semester. However, she dropped the second semester of chemistry because, as she said, "I didn't understand it." This teacher continues to teach third grade as her primary profession,

Mike Ciesielski

even though she is lacking in background knowledge of basic chemistry concepts. Incidents such as these make one wonder about how many other prospective elementary science teachers and other excellent students are being turned-off to chemistry because of the way it is taught!

National Science Education Standards

Two different groups in the United States have produced standards for teaching children chemistry in kindergarten through grade twelve. These are part of the *National Science Education Standards* (NSES) and were developed by chemistry experts under the direction of the American Chemical Society and published by the National Academy Press (National Research Council, 1996) and the American Association for the Advancement of Science (1993), *Benchmarks for Science Literacy*.

The recommendations of the NSES and the *Benchmarks* are quite similar. One major difference is that the recommendations of the NSES are less specific about the appropriateness of specific ideas and content at particular grade levels. Recommendations in the NSES are made for grades K–4, 5–8, and 9–12, whereas recommendations in the *Benchmarks* are made for K–2, 3–5, 6–8, and 9–12. In terms of chemistry instruction, the NSES for elementary students do not include the study of atomic theory, whereas this is included in *Benchmarks*.

A study by Liu (2006) using a U.S. sample from the Third International Mathematics and Science Study found that

> "…third-grade students were developing understanding on mixtures, and fourth-grade students were developing understanding on separating mixtures; seventh- and eighth-grade students were only at the beginning level of differentiating chemical properties from physical properties; they were not ready for the particulate model of chemical change."

The findings suggest that the *Benchmarks* and the *Atlas of Science Literacy* (American Association for the Advancement of Science, 2001), a resource for curriculum development based on the standards, may have *overestimated* the competences of elementary middle school and high school students. The NSES may be more realistic in terms of chemistry instruction.

The study by Liu supports this, as does a study by Harrison and Treagust (2002).

Another problem is that both sets of standards are considered to be *minimal*. That is, many states include *additional* standards in their state standards for elementary and middle school students. This has, in turn, caused science textbook publishers to increase the content of their textbooks to meet the state standards by publishing specific textbooks for given states. In chemistry, for example, one textbook publisher includes introducing atoms in grade three for children in Indiana.

Even though both sets of standards mentioned above have recommendations concerning what children in the various grade levels should learn, the *No Child Left Behind* legislation has, in reality, often compelled schools to ignore science, while spending much more time on reading and math skills, which are tested by a wide variety of both state and national tests. In the June 4, 2007, issue of *Time Magazine*, a feature article entitled "Report Card on No Child Left Behind" stated that "because the law holds schools accountable only in reading and math, there is growing evidence that schools are giving short shrift to other subjects." This makes it even more important that teachers and schools carefully choose what concepts in science they will teach across the different grade levels in order to make the instruction more effective.

The Reform of Chemistry Education

In the 1960s, reform of science education focused on the science process skills of observing, inferring, classifying, predicting, communicating, controlling variables, measuring, and doing experiments. These skills are extremely important and need to continue to be emphasized. And now, students are expected to give explanations of their observations.

Chemistry educators have known for many years that chemistry can be taught on three different levels: macroscopic, particle, and symbolic (Johnstone, 1990, 1993) (see Fig. 1). Although the levels are displayed on an equilateral triangle, teachers should not infer that the three levels are equally easy to understand, nor should be taught to students of all grade levels. Studying chemistry at the macroscopic level is the easiest level to comprehend because this level is less abstract and more accessible, depending extensively on the use of the senses (in particular, seeing, and perhaps smelling, hearing, or touching (with caution), but never tasting).

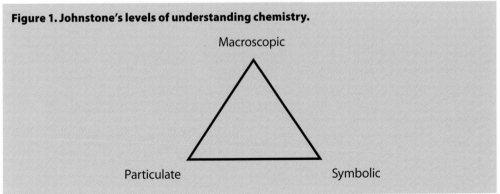

Figure 1. Johnstone's levels of understanding chemistry.

Macroscopic

Particulate   Symbolic

Used with permission from the *Journal of Chemical Education*, Vol. 70, No. 9, 1993, pp 701 © 1993, Division of Chemical Education, Inc.

The other two levels of Johnstone's triangle (particulate and symbolic) are considerably more abstract and are not developmentally appropriate for most children in grades K–4 (and even in grades 5 and 6, and for some children in grades 7–9.) When topics such as atoms are introduced in grade 3 (as they are in one textbook series approved for adoption in some states), many children will memorize definitions. Even some teachers are challenged to explain the difference between an atom and a molecule. The complexity of understanding chemistry at the particulate level is simply inappropriate for elementary children.

There are numerous activities and experiments that can be done on the macroscopic level, which children in grades K–8 find interesting and exciting. Some are even mathematics related. An excellent example of a reference for appropriate activities for grades 3–8 is *Inquiry in Action: Investigating Matter Through Inquiry* (ACS). Unfortunately, what has happened in the past several years is that much of the content that used to be taught in high school chemistry courses in the 1950s has been moved to middle school or lower, even though it is not included in the NSES or *Benchmarks* for middle school children. If all children in the United States mastered either set of these standards, they would be ready for high school science courses where more complex content is appropriate.

Research on Children's Understanding of Chemistry

What does the research tell us about students' understanding of chemistry? We examine children's understanding of chemistry at the high school, middle school, and elementary school levels. We have included selected findings about high school students because it seems reasonable to conclude that if students at the high school level do not understand certain concepts that are also being introduced in middle or elementary schools, then these same concepts will not be understood by younger children either. Although we have examined a large number of studies (42) on this topic, we limit this review to those studies that seemed to be most relevant, very informative, comprehensive, and well researched. Studies that were examined were not limited to students in the United States because it was assumed that children's developmental levels are generally consistent across countries.

Many studies over the past 50 years have focused on children's understanding of chemistry on the particle and symbolic levels. This review will include cross-age studies, as well as studies of high school students that provide information about student understanding of the particle nature of matter and symbolic representations. If high school students have difficulty learning these aspects of chemistry, it is reasonable to assume that this content is probably too complex for most middle and elementary school students.

High School Students' Understanding of Chemistry. Griffiths and Preston (1992) interviewed 30 grade 12 students drawn from 10 high schools in Newfoundland on concepts relating to molecules and atoms. Students were classified into three groups: academic science (with at least two high school science courses and an average of above 75% university-bound science students), academic nonscience (no science courses and overall average of at least 75%), and nonacademic (average less than 75% and not taking science courses).

Results of the study showed that there were 52 different misconceptions relating to atoms and molecules. Even the very best group had a considerable number of misconceptions. One-third or more of the sample shared 19 misconceptions. Some of the common misconceptions are listed in Table 1. If high school students of the top group in Newfoundland have these misconceptions, it follows that the misconceptions about atoms and molecules of the middle school students will be even more numerous. It is also likely that this same problem exists in the United States and other countries.

Another study that has implications for what chemistry concepts are appropriate for students of various grade levels was conducted by Abraham et al. (1994). They explored students' understanding of chemistry at grades 9, 11, and 12, and in an introductory college chemistry course and found that

Table 1. Misconceptions of Grade 12 Students About Atoms and Molecules

Student Idea	% Students
1. Water molecules are composed of two or more solid spheres	70
2. A water molecule is macro in size.	50
3. Water molecules within a phase may have different sizes.	40
4. Water molecules in ice touch each other continuously leaving no space.	50
5. Collisions may result in a change of atomic size.	50
6. All atoms are alive.	50
7. Atoms are alive because they move.	50

- "Both reasoning ability and experience with concepts account for the understanding of chemistry concepts.
- Students at all levels tended not to use atomic and molecular explanations for chemical phenomena. Although the use of atomic and molecular models increases with increased exposure to chemistry concepts, it is still low, even among college students.
- There were no predictable patterns in the frequency of alternative conceptions with respect to experience with the concept..."

Middle School Students' Understanding of Chemistry. A recent comprehensive study of middle school students' understanding of chemistry by Nakhleh et al. (2005) compared middle school students' ideas about matter with those of elementary students. They found that most of the middle school students interviewed knew that matter was composed of atoms and molecules. Some students were able to use this information to explain processes such as phase changes of water but that their knowledge frameworks were inconsistent because of their fragmented ideas. This points to the difficulty of assimilating the particle level acquired by instruction into their formerly acquired macroscopic knowledge.

Elementary School Students' Understanding of Chemistry. An earlier comprehensive study by Nakhleh and Samarapungavan (1999) explored elementary school children's beliefs about the particulate nature of matter before they had any formal instruction on this topic. Fifteen students (ages 7–10) were interviewed concerning their understanding of the microscopic and macroscopic properties of solids, liquids, and gases, as well as their understanding of phase changes and dissolving. Sixty percent had macroparticulate beliefs; 20% had microparticulate beliefs, and 20% held macro-continuous beliefs about matter. The children's beliefs were not consistent across the variety of substances from continuous to particulate substances. An example of children's thinking is illustrated because Linda (age 8) "held a macroparticulate view of matter by stating that the substance was 'made of little pieces' or could be divided by human action." An excerpt of Linda's interview (Nakhleh and Samarapungavan, 1999, p. 787) is shown in Figure 2. Linda was classified as "macroparticulate" because her incomplete view of matter did not include that all of the particles were identical in size and shape, nor were they made of identical atoms. No student (ages 7–10) in the study had a completely accurate view of the composition of sugar.

Cross-Age Studies on Understanding Chemistry. As indicated earlier, Liu (2006) studied the competence levels in understanding the "matter concept" at grades 3 and 4, 7 and 8, and grade 12. Findings included:

- Third-grade children were beginning to recognize mixtures and to separate them.
- In grades 7 and 8, students had mastered recognizing mixtures.
- High school students had developed an understanding of molecular models and were beginning to understand atomic structure.

Figure 2. Elementary school child's thinking about particulate nature of matter.

Researcher	Now take a look at this sugar cube and say, is this just one big piece of material or is it made up of little bits?
Linda	*It's a little piece of sugar.*
Researcher	Okay, Now think about the smaller little pieces of sugar that this cube is made of. Uh, are they the same or different?
Linda	*Different.*
Researcher	Okay, can you tell me in what way they're different?
Linda	*They're all probably shaped different... I don't know how little they are. [Linda constructs Play-Doh models which are very small round and oval objects]*
Researcher	What shapes would these little pieces be, you think **they'd be all different** shapes?
Linda	*Kinda circle and kinda oval.*
Researcher	Okay [pause] You think that there would be any other shapes, like squares and triangles, or things like that?
Linda	*Yeah [long pause].*
Researcher	You know, this tiny little piece of sugar, uh, What would they taste like?
Linda	*Sweet.*
Researcher	Sweet. What color would they be?
Linda	*White. White and [pause] I dunno. White.*

Used with permission from the *Journal of Research in Science Teaching*, Vol 36, No. 7, pp 787; © 1999, Wiley-Liss, Inc., a subsidiary of John Wiley & Sons, Inc.

Mike Ciesielski

An additional research report of Liu and Lesniak (2006) provides more details about the study in general. Cross-age studies of the matter concept, particularly when the same authors conduct the studies are extremely useful in determining the appropriate level at which chemistry concepts should be taught.

Conclusions and Recommendations

In an unpublished study now being prepared for publication by D. Gabel, L. Cardellini, and L. Wozniewski, data were collected on college students' understanding of three sets of concept-pairs, prior to and after taking an introductory college chemistry course in both the United States and Italy. The pairs were chemical vs. physical change, burning vs. decomposition, and melting vs. dissolving. All of these topics are included in the chemistry textbooks examined at the middle school level. The test contained a macroscopic, a particulate, and a symbolic question on each of the six concepts. Results indicated that there was no significant improvement by college chemistry students after one year of college chemistry. The average pretest score was 7.5, and average posttest score was 9.0 out of a possible 18. This indicates that if students do not learn about these everyday concepts before they take a high school or college chemistry course, they may never learn them! The above topics are appropriate for instruction at the upper elementary and middle school levels. The macroscopic level would be suitable for all children. The age level for the introduction of the particle and symbolic levels is questionable and will depend on the developmental level of the child. Presenting these concepts before students are developmentally ready is likely to turn students off from chemistry, rather than making them fans.

Because of this great diversity in chemistry instruction at the elementary and the middle school levels, as can be seen from the variety of textbooks and programs in use, it is very difficult to know exactly what chemistry concepts students understand when they enter high school. Of even greater concern, is whether students have had a positive experience when studying chemistry so that they are looking forward to taking high school chemistry and perhaps will even consider majoring in chemistry in college. As indicated by R. L. Wolke,

> "The funnel that leads to chemical careers can have a very wide top. To collect more chemists at the bottom, we must pour more young science fans, not just chemistry fans, into the top, through the funnel. As more and more children are turned on to a broad array of natural wonders, they will sort themselves out while passing through the educational funnel, and a fair proportion will inevitably wind up in the chemical sciences."

Let's hope that he is right, and give this a try!

Recommended Readings

Keeley, P.; Eberle, F.; Farrin, L. *Uncovering Student Ideas in Science: 25 Formative Assessment Probes*, NSTA Press: Washington, DC, 2005. This book is very helpful in determining what high school students actually know.

Kessler, J.; Galvan, P. *Inquiry in Action: Investigating Matter Through Inquiry*, 2nd ed.; American Chemical Society: Washington, DC, 2005. This is an excellent reference for both activities for grades 3–8, as well as information about guided inquiry teaching techniques.

Linse, P. L.; Licht, P.; deVos, W.; Waarlo, A. J. *Relating Macroscopic Phenomena to Microscopic Particles*. CD-B Press: Utrecht, The Netherlands, Center for Science & Mathematics Education, University of Utrecht, 1990. This book contains the proceedings of a seminar held at the University of Utrecht. It contains seven plenary lectures and 19 invited papers, most of which are related to the teaching of chemistry at the macroscopic and microscopic levels. This book may be available from the Center for Science and Mathematics Education, University of Utrecht, P.O. Box 80.008, 3508 TA Utrecht, The Netherlands.

Liu, X.; Lesniak, K. Progression in children's understanding of the matter concept from elementary to high school. *J. Res. Sci. Teaching* 2006, *43*, 320–347. This very comprehensive study of children's understanding of chemistry consists of interviewing students from grade K–8 and grade 10 on the matter concept. The study was carefully planned and included in a pilot study. The general conclusions were that children progress from perceptual characteristics and uses and benefits to perceiving physical properties and change in grade 4 and up, to perceiving chemical properties and change in grade 5, chemical changes in grade 6, and finally to perceiving the particulate model of matter in grade 10.

Mike Ciesielski

Recommended Web Sites

http://www.chemistry.org/kids (accessed March 24, 2008). An excellent Web site sponsored by the American Chemical Society with a wide range of suggested activities for various grade levels.

http://www.stevespangler.com (accessed March 24, 2008). A Web site that was recommended by several teachers of all levels that has easy-to-use activities for all areas of science.

References

Abraham, M. R.; Williamson, V. M.; Westbrook, S. L. A Cross-Age Study of the Understanding of Five Chemistry Concepts. *J. Res. Sci. Teaching* 1994, *31*, 147–165.

American Association for the Advancement of Science (AAAS). *Atlas of Science Literacy;* National Science Teachers Association: Washington, DC, 2001.

AAAS. *Benchmarks for Science Literacy;* Oxford University Press: New York, 1993.

Griffiths, A. K.; Preston, K. R. Grade 12 Students' Misconceptions Relating to Fundamental Characteristics of Atoms and Molecules. *J. Res. Sci. Teaching* 1992, *29*, 611–628.

Harrison, A. G.; Treagust, D. F. The Particulate Nature of Matter: Challenges in Understanding the Submicroscopic World. In *Chemical Education: Toward Research-Based Practice*; Gilbert, J. K., Jong, O.D., Treagust, D. F., Van Driel J. H., Eds. Kluwer: Dordrecht, 2002; pp 189–212.

Johnstone, A. H. *Fashion, Fads, and Facts*. Paper presented at the American Chemistry Society Meeting, Washington, DC, September 1990.

Johnstone, A. H. The Development of Chemistry Teaching: A Changing Response to Changing Demand. *J. Chem. Ed.* 1993, *70*, 701–705.

Liu, X. Student Competence in Understanding the Matter Concept and Its Implications for Science Curriculum Standards. *School Sci. Math.* 2006, *106*, 220–227.

Liu, X.; Lesniak, K. Progression in Children's Understanding of the Matter Concept from Elementary to High School. *J. Res. Sci. Teaching*, 2006, *43*, 320–347.

Nakhleh, M.; Samarapungavan, A. Elementary school children's beliefs about matter. *J. Res. Sci. Teaching* 1999, *36*, 777–805.

Nakhleh, M., Samarapungavan, A.; Saglam, Y. Middle school students' beliefs about matter. *J. Res. Sci. Teaching* 2005, *42*, 581–612.

National Research Council. *National Science Education Standards;* National Academy of Science: Washington, DC, 1996.

Wolke, R. L. Forget chemistry: Letter to the Editor. *Chem. Eng. News* 2006, *36*, 3.

Chapter 15

Using the Research-Based Standards To Help Students Learn

by Diane M. Bunce, Sharon Hillery, and Elena Pisciotta

Diane M. Bunce *earned her B.S. in chemistry from LeMoyne College in Syracuse, NY, an M.A.T. from Cornell University, and a Ph.D. in chemical education from the University of Maryland (Advisors: Marge Gardner and Henry Heikkinen). She taught high school chemistry for six years in New York, North Carolina, and Maryland before returning to graduate school at the University of Maryland. She served as one of the original authors on three of the ACS curriculum projects (ChemCom, Chemistry in Context, and Chemistry). She is the founding editor of the chemical education research feature in the* Journal of Chemical Education. *Diane is currently professor of chemistry at The Catholic University of America. Contact e-mail: bunce@cua.edu*

Sharon Hillery *received a master's degree in organic chemistry from the University of Virginia. She has taught at the community college level and, for the past 22 years, has been a high school chemistry teacher. Twenty of those 22 years were spent at Georgetown Visitation Preparatory School in Washington, D.C., where she was Science Department chair for 10 years. She is currently on staff at Maret School in Washington. Contact e-mail: sharon.hillery@ maret.org*

Elena Pisciotta *earned her undergraduate and graduate degrees in chemistry education from the University of Maryland, College Park. She began her teaching career in 1968, took a 20-year hiatus to raise her three daughters and most recently returned to full-time teaching in 1992. Elena co-chaired the High School Teacher Day at the 2005 National ACS Meeting in Washington, D.C. and has won the ACS Mid-Atlantic Regional Meeting award for High School Teachers. Currently, she is teaching AP Chemistry at Damascus High School in Montgomery County, Maryland. Contact e-mail: Elena_S_Pisciotta@mcpsmd.org*

Introduction

Many students experience trouble with at least some concepts in a high school chemistry course. Students often explain their difficulty as a consequence of either not being "good" at math or science. Sometimes students place the blame on the teacher as someone who knows the science, but is not skillful at teaching. Whatever the problem, the thought of taking a chemistry course can cause a good deal of anxiety in many students. So why is chemistry perceived by some students as being so difficult? Is the subject inherently complex and accessible to only the top academic tier of students? If this is true, then why do the *National Science Education Standards* (NSES) include understanding of chemistry and chemistry-based principles as necessary for a quality education? Research suggests several reasons why chemistry is perceived as difficult, including the mismatch that can occur between the way chemistry is taught and how students learn. It is our premise, and that of other researchers, that chemistry is something that all students can learn if it is taught in a way that is aligned with how the brain operates. This chapter will explore some current ideas in the research of learning and teaching and then attempt to integrate this research with the NSES. We also offer some practical teaching examples keyed to the pertinent NSES to facilitate student learning.

Research in Learning

Researchers' analyses of teaching and learning methodologies have identified several basic principles involved in the teaching and understanding of subjects such as chemistry. Included in their findings are the following observations:

- In order for new knowledge to be learned, it must be integrated with what the learner already knows. This means that it is crucial that both the student and teacher ascertain what the student knows about a topic before new information is presented (Novak and Gowin, 1984).
- There is a difference between truly understanding the subject matter and simply being able to pass a course. This difference can be highlighted by asking students to apply what they know to novel situations rather than using problems that are parallel to what has been shown in class (Novak and Gowin, 1984).
- Students must be actively engaged in the learning process. They cannot be passive recipients of knowledge. It is essential that they are involved in the process of integrating new knowledge with the knowledge they already have (Bodner and Klobuchar, 2001; Nurrenbern, 2001). One way to actively involve students is to engage them in group work where answers to problems are debated and agreed upon by a group of peers (Johnson, Johnson, and Smith, 1998).

39th IChO

- Students need time and opportunities to *interpret* chemical concepts in their own work, *practice* applying these concepts to new situations, and *reflect and modify* their understanding of chemical concepts (Karplus, 1980).
- In order for students to be able to reliably learn new concepts, they must be aware of the process in which they are engaged when they integrate new information (Flavell, 1979). This self-awareness, known as *metacognition*, will aid in the process of learning a specific concept and help in understanding future related concepts.
- Chemistry is one of the first courses students encounter in science that has a large number of abstract concepts. Students can see muscles and bones in biology, and minerals and rocks in geology, but atoms, ions, and molecules are not visible to the naked eye. Molecular interactions are central to understanding chemistry. This move from concrete, observable phenomena to abstract, unseen molecular interactions is not always easy for students at different levels of cognitive maturity (Nurrenbern, 2001).
- Chemists move seamlessly among the macroscopic (large-scale phenomena), particulate (interactions of atoms, ions, and molecules), and symbolic (chemical formulas and mathematical equations) levels. Unless students are explicitly taught to link these three views of matter and move back and forth among them, they will not be able to understand chemical concepts the way chemists do (Johnstone, 1997). (See chapter 14 for a more detailed discussion of Johnstone's levels.)
- The short-term memory of the brain has a limited capacity and can easily be overwhelmed by new concepts explained with a new vocabulary and using unfamiliar equipment and techniques (Simon, 1979).

Research in Teaching

Research has also elucidated several aspects of teaching that can be used to help students learn effectively, including the following:

- "One size" teaching does not fit all students. There are students, who either because of their previous knowledge or their preferred mode of learning, will learn most effectively from a lecture format. However, these students are few and far between. Even if lecture is the preferred mode of learning for students, it may not be effective at times in the student's life when there are competing events for the student's attention. The teacher can help all students by presenting different topics in different formats and/or making alternative formats available to students (Campbell et al., 2001). These alternative formats may include animations on the Web, lab experiments, demonstrations, small group work, research papers, free response answers that require logical arguments, structured notes, diagrams, and model kits.
- The role of a teacher has changed from the "keeper of the knowledge" to the "designer of a learning environment" where the student can operate successfully. This idea of the teacher as a designer of the learning environment is a role that is deeply imprinted on the human psyche. Parents of toddlers (humans new to the concepts of life) baby-proof their homes so that the children do not hurt themselves. They also create learning opportunities that match the abilities of the child and enable the child to master and then grow into new applications of skills by providing age-appropriate toys, tasks, books, CD/DVDs and other learning materials. The teacher's role as a learning environment designer is similar. The teacher cannot learn the chemistry for the student. However, the teacher can construct the activities and materials that will allow students to confront, practice, reflect, and modify their learning experiences (Karplus, 1980).
- If students operate in a learning environment where their egos are protected from undue stress, their naïve ideas listened to and gently critiqued with new directions provided, students, like all human beings, will have a better chance to grow in their understanding (Novak and Gowin, 1984).
- Students need assessments that truly help their understanding. Tests and other forms of student work should not be viewed as barriers to be overcome. Rather they should be considered signposts to help direct student learning. This doesn't mean that students should not be given a grade for accomplishing the task of understanding, but rather that the assessment should be well focused and seen as a learning opportunity in and of itself (Fuchs and Fuchs, 1986).

Learning Research Goals and the *National Science Education Standards*

Many of the research results on learning and teaching are reflected in the NSES. It is the goal of this chapter to offer suggestions on how to implement the results of research on learning and teaching, as incorporated into the NSES to plausible and effective activities within the classroom. In keeping with this goal and as a result of discussions of problems, we have encountered in our own classrooms, we will address four basic research-based learning and teaching goals:

- Engage students and encourage them to take responsibility for their own learning
- Provide students with a voice in their own learning
- Approach the learner as an individual
- Apply chemistry concepts to the real world

Each of these goals will be discussed in terms of the relevant learning/teaching theory, connections to the relevant NSES, practical classroom approaches to addressing the goals/standards, and the enhanced delivery of these approaches through the use of technology. Table 1 presents an overview of this approach.

Additional resources are available to teachers who wish to address the NSES but are unsure how to approach this daunting task. Textbooks such as the *World of Chemistry* (Zumdahl, Zumdahl, and DeCoste, 2006) and others include references to the standards in their supplemental materials.

Table 1. Overview of Learning Research Goals and the NSES

Goal	Theory	NSES	Classroom Approach	Technology
Engage students/ Encourage responsibility for their learning	• Constructivism • Cooperative learning • Learning cycle	Teaching Standards A, B, D	• Cooperative learning groups • POGIL	• Interactive Web resources • Student response systems (Clickers) • ConcepTests
Provide students with a voice in their learning	• Metacognition • Reflective learning	Teaching Standards B, D, E	• Formative evaluations • Surveys • Class boards	• Class Web site • Clickers
Learner as an individual	• Learning styles • Piaget • Previous knowledge (Ausubel)	Teaching Standard B	• ConcepTests • Surveys • Formative evaluation	• Clickers • Class Web site • Interactive computer testing • Computer animations
Apply chemistry to real world	• Learning cycle • Three views of matter • Meaningful vs. rote learning (Ausubel)	Content Standard G	• Inquiry laboratory experiments • Web site or literature research of current topics	• Web sites • Computer searching of literature

*POGIL, Process Oriented Guided Inquiry Learning.

The third column of Table 1 connects these learning research goals to specific standards. The NSES Teaching Standards are structured so that teachers can select, adapt, and design content that emphasize inquiry learning for students (Teaching Standard A). Teachers should function as facilitators of learning by designing opportunities for student discourse about science and by encouraging students to take responsibility for their own learning (Teaching Standard B). Learning science requires tools, materials, media, and technology for students to engage in hands-on exploration; teachers are the architects of designing these learning environments (Teaching Standard D). Teachers should develop communities of learners to reflect the values and processes of inquiry (Teaching Standard E) and engage students and encourage them to take responsibility for their own learning.

Engage Students and Encourage Them To Take Responsibility for Their Own Learning

Theory. When students are part of a cooperative group, the group itself encourages each member to contribute his or her part. Membership in a group implies joint responsibilities, as well as joint benefits. Attendance in class can improve if each group member has a specific role to play in the cooperative group such as leader, facilitator, reporter, or presenter. The group itself is diminished if a member is not present and someone else must assume the responsibility of the absent member. This scenario often prompts the group to apply pressure on its members to be present in class and to fulfill their responsibilities to the group. As a group member, each student is an integral part of the learning experience and not just a faceless member of the larger class. Learning takes place through the group's discourse. Usually, this discourse goes through several stages from checking the parameters of the problem to making sure that the solution is logical (Daubenmire, 2004).

Students must construct their own knowledge as opposed to being passive recipients of the teacher's information. Students should be provided an opportunity to develop a concept, apply it, reflect on it, and modify it.

Corresponding National Science Standards (Center for Science, 1996) (Table 1)

Teaching Standard A

Teachers of science plan an inquiry-based science program for their students. In doing this, teachers

- Select science content and adapt and design curricula to meet the interests, knowledge, understanding, abilities, and experiences of students.

Teaching Standard B

Teachers of science guide and facilitate learning. In doing this, teachers

- focus and support inquiries while interacting with students,
- orchestrate discourse among students about scientific ideas, and
- challenge students to accept and share responsibility for their own learning.

Teaching Standard D

Teachers of science design and manage learning environments that provide students with the time, space, and resources needed for learning science. In doing this, teachers

- make the available science tools, materials, media, and technological resources accessible to students and
- identify and use resources outside the school.

Classroom application—Cooperative learning. When Advanced Placement chemistry classes are taught using guided inquiry techniques, students can become much more independent learners. One way to use guided inquiry is through use of Process-Oriented Guided Inquiry Learning (POGIL) curricula materials (POGIL, 2007). For instance, in the POGIL Activities for General Chemistry (Moog and Farrell, 2006), students move from critical thinking questions about the information that supports the electron shell model of atomic structure to exercises (application questions) and problems (extension questions) using that model. Learning takes place as students work in small groups and discuss and debate the answers and the reasons for them with each other. Students learn to rely on the group to interpret the material before asking the instructor for help with questions that they don't immediately understand. The basis of group assignment differs among teachers but can include mixed ability groups (one high ability, one low ability, and two middle ability students) or single ability (four high ability or four low ability students). Some teachers find that putting the lowest-ability students together in a group forces them to draw on their own knowledge and not depend on the "smart" kids for the answers. (Chapter 4 discusses POGIL in detail.)

Cooperative learning can also be accomplished by using other published cooperative learning activities such as Team Learning Sheets (Zumdahl, Zumdahl, and DeCoste, 2002b). These sheets are designed to be used in class to encourage active learning. In the laboratory, cooperative learning groups can use Report Sheets (Zumdahl, Zumdahl, and DeCoste, 2002a) to gather and analyze data. For example, an inquiry lab on solubilities can help clarify the idea of insoluble versus soluble ionic compounds. Students are given a number of unknowns and asked to plan their strategy before attempting to identify the unknowns through measurement of conductivity, pH, reactivity, and solubility, as different unknowns are mixed together (Zumdahl et al., 2002a).

Classroom application—Use of alternative or supplemental materials. Providing students with an outline of the topics that will be discussed in class and pertinent concepts that can be expanded during the discussion along with space to work out examples can help students stay engaged and take responsibility for their own learning. Projecting an outline of what will be covered in class, while class notes are discussed and elaborated, enables the teacher to face the class and engage students more directly. Problems can be worked out with the class as a whole, in small groups, or as individual assignments. The results are then discussed, projected, and added to the notes' outline.

Alternative approaches to engaging the learner include interactive demonstrations and simulations. One such simulation deals with the underlying principle of Rutherford's experiment. In this simulation, an unknown geometric shape is determined through an analysis of the path marbles take (angles of deflection) when passed under a board that obscures the identity of the geometric shape (triangle, pentagon, circle, etc.) (Zumdahl et al., 2002a). Just as Rutherford could not see the atom, students cannot see the geometric shape. Rolling marbles under the board and observing the deflection patterns allow the students to predict the shape, which is analogous to the procedure Rutherford used to hypothesize about the atom. This simulation helps to engage students in the learning process and shows them how discoveries can be made on the basis of logical analysis of appropriate data, even though a phenomenon cannot be viewed directly.

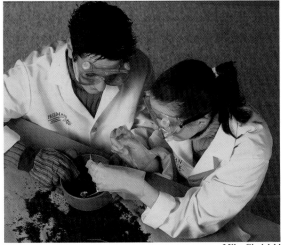

Mike Ciesielski

Supplemental information for concepts can be listed on a teacher's Web page or course Web site. Another way to introduce supplemental material into the learning environment is through appropriate posters and other visual information in the classroom. For instance, a good way to start a discussion of the mole is with the *ChemMatters* poster (Soule, Schwartz, and Pryde, 1985) that illustrates a mole each of marbles, pennies, and moles (animals), as well as other examples. This visual representation of an abstract concept helps students comprehend the size of a mole.

Supplemental information for lab experiences can include online simulated lab activities that reinforce specific concepts or lab procedures.

Technology. Class Web sites that provide students with alternative or supplemental resources can help empower them to take responsibility for their own learning. Computer files or URL's supporting concepts taught in class can be an effective way to both engage students and help them become responsible for their own learning.

Class Web sites also provide the ability to post class messages when necessary. An example is communication during inclement weather when it is important not to lose the learning opportunity interfered with by the weather by letting students know what they should be working on. As high schools become more Web-oriented, many additional files can be posted for the students' benefit, including answer keys to worksheets, students' grades (with individual password-protected access), discussion groups, assignments, writing guidelines for lab reports, and course syllabus. Students will then be better able to take charge of their learning. (See chapter 7 for a detailed discussion of how technology can improve student learning in chemistry.)

Provide Students With a Voice in Their Own Learning

Theory. Students and teachers are both stakeholders in the learning outcomes of the classroom. Since both are involved in the process, both should have input. This input does not necessarily mean that the teacher and students are equal partners in determining what will be taught or how much content will be covered in a school year. But by listening to students, teachers can better understand whether the approach they are using is meeting their students' needs. Students can learn that there is a difference between "whining" and actively engaging in a constructive discussion of what is or is not working in their quest for true understanding. If students share the responsibility for the learning environment, they are more likely to share responsibility for their own learning.

Corresponding National Science Standards (Table 1)
Teaching Standard B, D, and E
- Teachers of science develop communities of science learners that reflect the intellectual rigor of scientific inquiry and the attitudes and social values conducive to science learning.
- Students are enabled to have a significant voice in decisions about the content and context of their work and require students to take responsibility for the learning of all members of the community.

Classroom application—Learning from their mistakes. One approach to helping students take responsibility for their learning is to provide students with time either in or outside class to discuss their errors and write corrections with reflections on how to avoid making the same mistakes in the future. This activity helps students internalize the concept and develop a strategy for avoiding repeated mistakes. Students can submit their corrections for credit.

Students are given some choice in terms of alternative evaluations. A science fair project/ class presentation or a fourth-quarter research paper/class presentations are options students might choose in place of a test.

Classroom application—Evaluation. Lab reports can be used as a means to evaluate more than a specific laboratory experiment's results. The conclusion section of a lab report can include both conclusions based on data and an evaluation of the lab. Such evaluations can contain answers to questions such as what worked, what didn't, was the lab objective achieved, and what was learned by doing the lab. If the student misses the intent of the lab, as evidenced by the answers to these questions, then the teacher can help guide the student to an understanding of the lab's purpose.

Students can take an active role in designing their learning environment by completing course evaluations throughout the course. The evaluations can include questions related to length of assignments, pace of course, use of lab work, and supplemental materials (videos, worksheets, etc.). Instead of learning that students don't understand a concept on a test or quiz, a regular process for determining how effective specific teaching approaches have been and possible recommendations for how to improve them is provided. More frequent evaluation or input from students enables teachers to change or modify their methods of teaching, while the course is still in progress.

Classroom application—Class boards. To encourage meaningful discussion between the members of the class and the teacher on issues of student learning, class boards can be set up. Elected student representatives to the board meet with the teacher on a regular basis (weekly or monthly) to discuss issues that are of importance to members of the class. Minutes of the meeting, including responses to the concerns raised, are posted on the course Web site and made available to all students in the class.

Technology. Class Web sites can be used to host online discussions (either in real time or in threaded discussion boards) where students and teachers discuss elements of the course that are of mutual concern. Class Web sites can also be used to post minutes from Class board meetings so that they can be easily accessed by all members of the class. Student response systems (Clickers) can be used for short, informal surveys of important, time-sensitive issues such as rescheduling a test due to an unexpected snow day.

Approaching the Learner as an Individual

Theory. Contrary to the typical reference of teaching a class, teachers do not actually teach a "class." We teach individual students. These students are at different stages of psychological development, academic and social maturity, attention level, assimilation into American society, emotional states, or have preferred approaches to learning (visual, verbal, or kinesthetic). All of these differences cannot possibly be addressed in each teaching activity, but teachers can provide alternative approaches to learning concepts, including those that focus on visual, oral, or kinesthetic approaches. Teachers can help teach students how to be focused and persevere in their studies when their lives are in turmoil— whether real or imagined. To do this, the learning activities should be designed to focus on the students and engage them in an interactive experience of the concepts whether through simulation, demonstration, laboratory, guided activity, collaborative groups, or problem solving. Only then do teachers have a chance to effectively reach a diverse audience. Students should be engaged in activities that make them aware of how much they understand a concept and what they still need to work on. This metacognitive process involves student reflection on the state of their knowledge and an assessment of what they know and what they need to know. When students become aware of their own internal cognitive processes, they can more closely monitor the progress of their learning and eventually become a more active participant in their learning.

Mike Ciesielski

Corresponding National Science Standards (Table 1)
Teaching Standard B
Teachers of science guide and facilitate learning. In doing this, teachers
- challenge students to accept and share responsibility for their own learning and
- recognize and respond to student diversity and encourage all students to participate fully in science learning.

Classroom application—Assignment of groups. Group activities can help students of different cultures more easily become incorporated into the class by acknowledging their contributions within a smaller group. It is important not to isolate a shy member of a minority group by placing him or her alone in a group that will not automatically solicit the student's input. Assigning two students of the same culture to a group of four provides a certain amount of support for all students. Groups can be changed on a regular basis to help each student learn to work with others and not become dependent on certain individuals in the original group. Group assignment is not the only thing to consider in this situation. It is important for the teacher to encourage full participation by all members of the cooperative group.

Classroom application—Addressing different learning preferences. A variety of teaching methods can be used to reach the auditory learner, the visual learner, and the student that is learning disabled (LD). The use of class notes that are expanded as the class discussion proceeds helps all three types of learners. This technique has also been used to aid learners who have been identified by professionals as needing a "note-taking buddy". In expandable notes, a basic outline of what will be covered is given to each student. As the class progresses through a discussion of the material, examples are added utilizing a conventional or tablet-type computer and LCD projector. Students can then add these examples to their outline. The example is discussed, and the students annotate their notes. The notes and examples are then saved on the computer and made available to students who were absent for that presentation or anyone else who wishes to check the quality of their notes against those of the teacher. Teachers are able to facilitate improved class participation and be more aware when a student seems confused or frustrated. In such a case, the student can be helped individually by the teacher at an appropriate time. Teachers can also help students focus more effectively on the concept being discussed through the class interaction.

Classroom application—Labs. Lab experiments allow students to view and manipulate on the macroscopic level what they have only read or learned about in class. Small groups of two or three provide an opportunity for all students to have a hands-on experience with a new concept. The ability to link the macroscopic with the symbolic view of chemistry is a major step forward in understanding. Once students are comfortable moving between these two views of matter, using a third view, particulate, to explain chemical phenomenon will likely be easier for them to grasp.

Technology. Use of Web pages as a repository for supplemental resources that are more visual than those used in class or as a file manager for class notes enables students to access the resources they need to help support their learning preferences or to supplement their understanding. Using technology in the classroom whether it is a tablet-style computer that supports free-form writing or computer-LCD projector combination as a way to both project the current information and to retain an electronic record, helps students who require additional support in the learning process.

Applying Chemistry to the Real World

Theory. In order for meaningful learning to take place, the student must integrate the newly acquired knowledge with that he or she currently possesses. If the new knowledge deals only with issues that the student encounters within school and appears to have little or no connection with the real world, there is a risk that the school knowledge, even if it is learned, will be held in isolation from real-world knowledge. This parallel processing of knowledge will make it difficult for the student to apply science concepts outside of a school setting, thus limiting the impact those scientific concepts have on the student's understanding of the world.

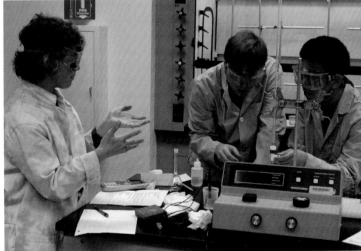

David Armer, USAFA

Corresponding National Science Standards (Table 1)
Content Standard G
    As a result of activities in grades 9–12, all students should develop an understanding of
    • science as a human endeavor and
    • the nature of scientific knowledge.

Classroom Application–Connecting life situations with chemistry concepts.
What better way to capture students' attention than by talking about food? Food is a natural blend of real-world concerns and chemistry. Physical and chemical changes can be illustrated

by calling upon the students' own experiences with food, including what happens to food left too long on the counter versus the refrigerator. Gas laws can be introduced with the opening of warm and cold soda bottles. Acid-base reactions can be addressed through the need for antacid tablets after a large meal. Even a discussion of pH can begin with the pool testing kits used by home or community pool lifeguards. This discussion can be expanded to include students' knowledge of aquarium pH balance or the advantages of using a pH-balanced shampoo. Limiting reagents can be addressed through the effect of running out of specific ingredients while cooking. Once students are hooked on the real-world topic, it is easier to direct their interest in the underlying chemical concepts.

The application of chemical concepts to real-world phenomena can be further investigated through research papers or written assignments. Use of Web-based search engines to locate print and electronic information is a worthwhile skill for students to develop while they are learning chemistry. LexisNexis (Elsevier, 2006) and EBSCO (Ebsco, 2007) are two search engines that many students are already familiar with. This Web-searching skill can be tapped to help locate sound scientific knowledge about real-world topics.

Technology. Using the Web as a source of information to link chemistry concepts with real-world phenomena is an effective way to research the underlying chemical concepts for phenomena outside the chemistry classroom. Learning how to locate information and judge its quality is an important skill for students. Some schools have developed their own guidelines for effective electronic searching and posted them on their school Web sites (Georgetown-Visitation, 2006). Other teachers have developed these guidelines within the structure of the chemistry class itself.

Conclusion

"Teaching to the Standards" may be misinterpreted as a way of interfering with the teacher's prerogative to respond to the needs of the students. However, the national standards are prescriptive and not restrictive. They encourage teachers to develop ways, as supported by research, on how teaching and learning take place, to meet the needs of their students. The proper use of the standards includes a recognition that they deal with the theory and research results of learning and teaching. To use the standards effectively, time is needed to analyze one's teaching in light of this external set of standards, but this can be time well spent. Teaching is a powerful tool that can either help or hinder student learning. Don't students deserve the best efforts, reflection, evaluation, and revision of our teaching methods and materials that we can provide? Emphasis on the student as central to the process of learning is at the heart of the standards. If the opportunity to create the most effective learning environment for students is not central to our role of teachers, then what is?

Recommended Readings

Moog, R. S.; Farrell, J. J. *Chemistry: A Guided Inquiry;* John Wiley & Sons: Hoboken, NJ, 2006; Vol. 3. The POGIL process promotes student engagement and involvement in the chemistry classroom. This text provides the activities to be used by the students working in groups that support the POGIL process.

Zumdahl, S. S.; Zumdahl, S.; DeCoste, D. *World of Chemistry.* Boston, MA: Houghton Mifflin, 2006. This high school textbook is very helpful if a teacher needs to show that the standards are being met. In the teacher's edition of the textbook, there are chapter planning guides for each chapter. The standards that apply to each section of the chapter are cited in this planning guide.

Recommended Web Sites

Georgetown-Visitation. *Evaluating Web sites*, from http://www.visi.org/news/detail.aspx?p ageaction=ViewSinglePublic&LinkID=1001&ModuleID=123&NEWSPID=1 (accessed March 18, 2008). This web site provides a detailed description of the criteria a student should use to evaluate a Web site. Having used the criteria, students can then decide whether a Web site is one they would like to cite.

POGIL, *Process-Oriented Guided Inquiry Learning*, http://www.pogil.org (accessed March 18, 2008). This Web site describes a National Science Foundation-funded approach to teaching chemistry through the use of guided inquiry worksheets with small groups. The curriculum section of the Web site provides some sample activity sheets that teachers can use in their classes.

References

Bodner, G.; Klobuchar, M. The Many Forms of Constructivism. *J. Chem. Educ.* 2001, *78*, 1107.

Campbell, J., Smith, D., Boulton-Lewis, G., Brownlee, J., Burnett, P. C., Carrington, S., Purdie, N. Students' Perceptions of Teaching and Learning: The Influence of Students' Approaches to Learning and Teachers' Approaches to Teaching. *Teachers and Teaching: Theory and Practice,* 2001, *7*, 173–187.

Center for Science, Mathematics, and Engineering Education. *National Science Education Standards*. Washington, DC: National Academies Press, 1996.

Daubenmire, P. *A longitudinal investigation of student learning in general chemistry with the Guided Inquiry approach.* Ph.D. thesis, The Catholic University of America, Washington, DC, 2004.

Ebsco. http://www.epnet.com/thisMarket.php?marketID=5 (accessed March 18, 2008).

Elsevier, R. *LexisNexis.* http://global.lexisnexis.com/us (accessed March 18, 2008).

Flavell, J. H. Metacognition and Cognitive Monitoring: A New Area of Cognitive-Developmental Inquiry. *American Psychologist* 1979, *34,* 906–911.

Fuchs, L. S.; Fuchs, D. Effects of Systematic Formative Evaluation: A Meta-Analysis. *Exceptional Children 53*, 199–208.

Georgetown-Visitation. *Evaluating Websites*, 2006. http://www.visi.org/news/index.asp?page action=ViewSinglePublic&LinkID=1001&ModuleID=123 (accessed March 18, 2008).

Johnson, D. W.; Johnson, R. T.; Smith, K. A. *Active Learning: Cooperation in the College Classroom.* Edina, MN: Interaction Book Company, 1998.

Johnstone, A. H. Chemistry Teaching—Science or Alchemy? *J. Chem. Educ.* 1997, *74,* 262–268.

Karplus, R. Teaching for the Development of Reasoning. *Res. Sci. Teaching* 1980, *10*, 1–9.

Moog, R. S.; Farrell, J. J. *Chemistry: A Guided Inquiry.* John Wiley & Sons: Hoboken, NJ, 2006; Vol 3.

Novak, J. D.; Gowin, D. B. *Learning How to Learn.* New York: Cambridge University Press, 1984.

Nurrrenbern, S. C. Piaget's Theory of Intellectual Development Revisited. *J. Chem. Educ.* 2001, *78*, 1107–1110.

Simon, H. A. Information Processing Models of Cognition. *Annu. Rev. Psychol.* 1979, *30,* 363–396.

Soule, R.; Schwartz, A. T.; Pryde, L. T. Mole Poster, *ChemMatters* 1987.

Zumdahl, S. S.; Zumdahl, S.; DeCoste, D. *Laboratory Experiments.* Boston, MA: Houghton Mifflin, 2002*a.*

Zumdahl, S. S.; Zumdahl, S.; DeCoste, D. *Teacher Resources Team Learning Sheets.* Boston, MA: Houghton Mifflin, 2002*b.*

Zumdahl, S. S.; Zumdahl, S.; DeCoste, D. *World of Chemistry.* Boston, MA: Houghton Mifflin, 2006.

Chapter 16

ACS and Its Role in the Future of Chemistry Education

by Steven Long and Mary Kirchhoff

Mary Kirchhoff is Director of the American Chemical Society Education Division. She earned her Ph.D. at the University of New Hampshire and taught at Trinity College in Washington, DC. Mary is a fellow of the American Association for the Advancement of Science and has worked with the Green Chemistry Program at the U.S. Environmental Protection Agency. Contact e-mail: m_kirchhoff@acs.org

Steven Long is a chemistry teacher and Science Department Chair at Rogers High School in Rogers, Arkansas. He earned his B.S.E. from the University of Arkansas–Fayetteville and an M.S. in Secondary Education from the University of Houston at Clear Lake. Steven has 32 years experience teaching chemistry, ChemCom, AP chemistry, and biology. He serves as a ChemCom Teacher Leader for ACS and has assisted with the ChemCom textbook. Contact e-mail: sjlong@rogers.k12.ar.us

Chemistry Education: Challenges and Opportunities

There is no shortage of experts when it comes to education. We have all been to school, and we know what worked for us—inspiring teachers, lots of homework, a rigorous curriculum, and caring parents. There is also no shortage of criticism when it comes to today's education system: teachers are poorly trained, students are lazy, classes are not challenging, and parents are too focused on their own careers to pay attention to their kids.

The challenge in education is that there is no single "right way" to do things. This is actually an opportunity though, as learning is a complex mixture of factors and a variety of approaches can be effective in meeting learning outcomes. The large number of variables involved in education presents a conundrum: how do we accurately assess what is effective in the classroom?

A tension exists in chemical education between covering traditional, in-depth content and introducing broader and more applied topics, such as industrial chemistry and sustainability. The "breadth vs. depth" argument is not new, but the increasing globalization of the chemical industry, the expanding multidisciplinary nature of science, and the growing emphasis on sustainability make a compelling case for change.

The American Chemical Society (ACS) directly supports the *National Science Education Standards* (National Research Council, 1996) in its *Statement on Science Education Policy* (American Chemical Society, 2007), and recommends implementing high standards:

> *Develop inquiry-based science curricula, based on content frameworks such as those provided by the National Science Education Standards (NSES) or AAAS Benchmarks, and include chemistry components at appropriate grade levels.*

The *National Science Education Standards* provide a framework for building a challenging science curriculum based on real-world interactions between students and the natural world. This chapter highlights some trends in chemistry education at the secondary and tertiary level, which influence the teaching of high school chemistry, and provides a snapshot of ACS resources that support the NSES.

K–12 Curriculum

Inquiry versus direct instruction. Public education has provided moral, cultural, and educational stability in the United States by weathering political debates and pedagogical experiments. In the years since the National Research Council published the NSES calling for changes in K–12 science content and pedagogy, the changes have been steady, but slow. This inertia has been seen in the cautious adoption of the important reforms suggested in the NSES, including the greater emphasis on integrated instruction and inquiry activities. In the first edition of *Chemistry in the National Science Education Standards* (American Chemical Society, 1997), authors challenged readers to place less emphasis on isolated science content and cookbook activities, and place greater emphasis on integrated, problem-based and inquiry activities.

JupiterImages

Adoption of the NSES approaches is making inroads. There are successful programs in which teachers are experimenting with inquiry and integrated, problem-based content and in which students are experiencing the joy and satisfaction of "doing science." However, the change has been challenging because of the requisite retraining of in-service teachers, the redesign of preservice teacher preparation programs, and the inherent inertia of the educational system.

Momentum is growing, and educators should persist in their efforts to reform and improve K–12 chemistry education. Direct instruction, including lecture, is not wrong, but it may not be the best technique for all instruction, just as inquiry may not be the optimal method for all laboratory activities. However, students need the opportunity to experience chemistry as it is typically encountered in the scientific world: an open-ended problem or question involving multiple disciplines, without a clear-cut solution, and with multiple possible solutions.

Diversity in student population. Another challenge—and opportunity—for K–12 chemistry instruction is the increasing diversity of the students in the classroom. The homogeneity that was typical of many schools, even 25 years ago, is rapidly disappearing. The increase in diversity certainly includes ethnicity, but it encompasses much more. Students today are more diverse in almost every aspect, including their range of cultural experiences, family structure, religion, and interests.

A teacher 25 years ago could use a teaching analogy in class based upon a television show from the previous night. There were typically three or four network channels, and popular shows were easily identified. Today's teacher would find it difficult to do the same with more than 100 television channels, satellite broadcasts, and Internet programming. This is one simple example of the diverse classroom that teachers face today. Expand this diversity to include cultural, religious, family/social, and ethnic diversity, and it is easy to see why the pedagogical bag of tricks that were effective in years past may have diminished results today.

Furthermore, a student's socioeconomic class can strongly influence his or her interest in science and science-related topics. A student whose family is struggling to make ends meet may have less interest in the science of global warming than a student from a more affluent background. Presenting examples that are contextually relevant to a diverse class engages students with the material in a more meaningful way but poses challenges for the teacher.

The increasing student diversity also represents opportunity. Students today need not fit into a predetermined mold. Instead of seeking jobs in the local area, students today compete in the global market. They know more, have greater access to knowledge and communication and will compete in the future for jobs that do not yet exist. Access to the Internet has widened the gap of knowledge and opportunity between the "haves" and the "have-nots". The opportunity is for students to capitalize on and create synergy from their diversity. Diversity can build strength and foster creativity; however, building this strength and creativity calls for innovation in the classroom on the part of educators.

The challenge of finding common experiences for effective instruction re-emphasizes the critical nature of the vertical alignment needed in science education. There must be a set of common science knowledge, skills, and activities that students have experienced upon which to build. The NSES provide a coherent backbone of content and skills if they are properly implemented. However, the quantity of knowledge and skills may need additional paring to distill it to a true common core. The educational debate over American education being an inch deep and a mile wide continues. (See chapter 13 for an expanded discussion on diversity in the chemistry classroom.)

Role of AP and IB courses. There is an increasing movement in the nation toward implementing Advanced Placement and International Baccalaureate programs (AP/IB). Schools seeking to prepare students for colleges and universities believe that the rigor of AP/IB courses is effective in propelling students into successful postsecondary studies. One recommendation emanating from the *Rising Above the Gathering Storm* report (The National Academies, 2007) focuses specifically on AP and IB courses:

> Enlarge the pipeline of students who are prepared to enter college and graduate with a degree in science, engineering, or mathematics by increasing the number of students who pass AP and IB science and mathematics courses.

AP and IB are the closest programs in the United States to providing a standardized national curriculum in chemistry. While there is great breadth of content and skills expected of students in either program, they are clearly defined for both the student and the teacher. The exams, projects, and scoring procedures allow the diverse student population to measure performance against each other and against a standard of excellence.

In addition, both AP and IB programs are recognized outside of the United States. The ability of students to attend college or university beyond our boundaries makes the AP/IB credits more valuable as students reduce global differences. Therefore, great care should be exercised in the design and the content of chemistry programs in both AP and IB. As secondary schools implement more AP/IB chemistry courses, there will be an increase in building Pre-AP and Pre-IB programs in middle and junior high schools to assure student success in the AP/IB courses. These Pre-AP and Pre-IB courses are intended to prepare every student, regardless of their background, to acquire the skills and knowledge to succeed in AP/IB chemistry. By default, the chemistry curriculum for many schools may be set by the College Board and the International Baccalaureate Organization. (See chapter 12 for a discussion of the redesign of the AP curriculum and exam.)

Undergraduate Education

Impact of globalization on professional preparation. The Executive Summary of *International Education and Foreign Languages: Keys to Securing America's Future* (O'Connell and Norwood, 2007) opens with the statement "A pervasive lack of knowledge about foreign cultures and foreign languages threatens the security of the United States as well as its ability to compete in the global marketplace and produce an informed citizenry." This observation is particularly relevant in light of the globalization of the chemical enterprise, a trend that suggests that today's students will be more likely to spend at least part of their careers working or doing business overseas. Consequently, it is increasingly important for students to have an international experience as part of their academic preparation (and to have studied a foreign language during the K–12 years). Science majors often have difficulty taking advantage of study abroad opportunities; however, this can disrupt the sequence of courses required for graduation, thereby prolonging time to degree.

International experiences for chemistry students are available and take a variety of forms. The new ACS-NSF-DAAD International Research Experience for Undergraduates program enables U.S. and German undergraduates to spend a summer conducting research in Germany and the United States, respectively. Some NSF-sponsored REU (Research Experiences for Undergraduates) programs, such as those offered by Syracuse University (Austria), the University of California, Santa Cruz (Thailand), and the University of Florida (France) are conducted overseas. Boston University takes another approach with the Dresden Science Program, designed for first-semester sophomores. Courses are taught in English (except for Intensive Beginning German, of course) and carry Boston University numbers, ensuring seamless transfer of credit. Students enrolled in the Dresden program take the courses they need in an international setting without disrupting the sequence of their major courses. Undergraduates benefit from these, and other, international opportunities, which serve to better prepare them for the global marketplace.

Graduate Education

Rise of interdisciplinary/multidisciplinary research. Many of the breakthroughs in science occur at the boundaries of disciplines, and the titles of new journals—*ACS Chemical Biology, ACS Nano*—bear witness to the increasing interdisciplinary nature of chemistry. Graduate and postdoctoral students should not only be encouraged to pursue research at the interface of disparate disciplines, they should also be provided opportunities to explore these intersections in venues outside of research. For example, both the European Union and the United States host annual summer schools to introduce students to green chemistry concepts and applications. These programs provide in-depth exposure to green chemistry, a topic that is frequently missing from the education of our students, as well as opportunities to present research and establish new collaborations. Similar programs in other multidisciplinary areas, such as molecular biology and nanotechnology, would allow graduate students to expand their knowledge and explore new research collaborations in interfacial fields.

Trends

Competitiveness initiatives. The alarm is sounding: Government leaders are concerned that the number of college graduates in science, technology, and engineering careers in the United States is declining. In chemistry, the 2005–2006 academic year actually saw record numbers of degrees awarded at both the undergraduate (11,938) and doctoral (2,321) levels. However, what is declining is the percentage of students earning degrees in technical fields, as more people are pursuing degrees in higher education. Furthermore, about one-third of Ph.D. graduates in chemistry are not U.S. citizens. As more attractive job opportunities open up in their native countries, these doctoral recipients are expected to return home in greater numbers, thereby creating a "brain drain" in the United States.

Both the U.S. government and American industries are concerned that the United States will lose its competitive edge in the world. Countries such as China and India are directing more resources to higher education in an effort to retain the best and brightest students. In addition, the off-shoring of high-quality, knowledge-intensive jobs in science and technology threatens the foundation of the U.S. economy, national security, and quality of life. *Rising Above the Gathering Storm* clearly outlines the status of U.S. science and technology competitiveness and presents recommendations for education, research, and economic policy. While our lives are increasingly global as described in *The World Is Flat* (Friedman, 2005), we wish to maintain our competitive edge through science and technology to ensure our American way of life.

Laboratory experiments, simulations, and online teaching tools. The evolving world of technology creates opportunities to help students understand chemistry. Rather than draw diagrams of a chemical reaction on a chalkboard or wave colored polystyrene balls in the air, teachers can use three-dimensional computer animations to transform the complexity of reactions into powerful visual images. Instead of a static ball-and-stick model of a water molecule, students can see the convoluted motions of the molecule and its bonds as a type of chemical ballet. Demonstrations that are too expensive or dangerous for typical classroom use are easily pictured in videos.

Mike Ciesielski

Today's students have the resources to visualize chemical processes better than those in the past. Despite these advances in technology, however, experiencing chemistry in a hands-on setting in the laboratory remains an essential component of chemistry education. The National Science Teachers Association notes that "For science to be taught properly and effectively, labs must be an integral part of the science curriculum" (National Science Teachers Association, 2007). The ACS requires that undergraduate students complete 500 laboratory contact hours for an approved program (ACS Committee on Professional Training, 2003), thereby reflecting the centrality of the laboratory experience to the preparation of professional chemists.

Most high schools have a laboratory requirement for chemistry courses. These requirements typically range from 20 to 40% of instructional time. However, as safety issues and the costs of chemicals and equipment continue to increase, schools are beginning to look at alternatives to hands-on experiments, a move that raises a number of questions: Should all laboratory work be replaced with simulations and technology? What is the role of the laboratory experience in chemistry? Can students learn chemistry as well or better on a computer?

Unless technology advances to the point where students can realistically manipulate equipment (rather than point-and-click) and sense changes such as odor and temperature (rather than read descriptions or gauges), simulations and Internet resources must remain supplemental to the hands-on laboratory experience. Chemistry teachers should embrace and use technology for the strengths it provides, but not expect it to be a panacea. (Chapter 7 provides a detailed discussion of the use of technology in chemistry education.)

Sustainability and green chemistry. While green chemistry began with a focus on higher education, the introduction of green chemistry at the high school level offers several benefits. First, green chemistry laboratories offer improved safety by minimizing the use of hazardous substances. Second, implementing greener laboratory experiments can decrease the amount of waste generated, thereby decreasing waste disposal costs. Third, green chemistry demonstrates that chemistry can be practiced in an environmentally responsible manner, which reinforces the interest in environmental issues demonstrated by many students.

Although the term "green chemistry" was introduced about 15 years ago, green chemistry concepts have not been widely incorporated into the curriculum. This is not surprising, as new ideas typically take time to get integrated into the mainstream. In addition, several barriers to incorporating these concepts exist: a relative lack of curricular materials, an already overcrowded curriculum, a perceived lack of rigor, and general inertia. Because inertia is a powerful force, a compelling case must be made for green chemistry, and the role of green chemistry in achieving sustainability makes a pretty strong case.

ACS collaborated with the Royal Society of Chemistry and the Gesellschaft Deutscher Chemiker to produce *Introduction to Green Chemistry* (Ryan and Tinnesand, 2002), a collection of six units that present key concepts in green chemistry. High school students are the target audience for this resource, which incorporates hands-on activities into the units, such as a Vitamin C clock reaction. The units in *Introduction to Green Chemistry* are aligned with the *National Science Education Standards.*

Additional curriculum materials continue to be developed. Several texts (Anastas and Warner, 1998; Clark and MacQuarrie, 2002; Lancaster, 2003) can be used for stand-alone green chemistry courses. Laboratory manuals (Doxsee and Hutchison, 2002; Kirchhoff and Ryan, 2002), primarily focused on organic chemistry, enable faculty members to introduce single experiments or an entire curriculum. The *Journal of Chemical Education* has published a number of green chemistry experiments through its "Topics in Green Chemistry" feature. Curricular materials are being disseminated through the University of Oregon's Greener Education Materials (GEMs) Web site (University of Oregon). This online resource allows educators to access, contribute, and review green chemistry curricular materials.

The Role of ACS and Advocacy

State and local control of K–12 education limits ACS's direct influence on a national level: an approval process, equivalent to the ACS approval process at the undergraduate level, simply does not exist for high school chemistry. Nonetheless, the ACS, through its Office of Legislative and Government Affairs (OLGA), is a strong advocate with the federal government for K–12 science education. ACS and the National Science Teachers Association lead the STEM (Science, Technology, Engineering, and Mathematics) Coalition, which promotes the critical need for resources for "K–gray" science education. The opt-in Legislative Action Network (LAN) enables ACS members to communicate directly with their legislators on issues related to STEM education, as well as pending legislation in other areas of interest to the science community.

Where ACS does shape precollege education is through its resources, which enable science teachers to provide the best education possible. Online resources, high school chemistry clubs, textbooks, *ChemMatters,* conferences and workshops, and national recognition all contribute to the vibrancy of the discipline. The following sections highlight a few of these valuable resources.

Chemical Education Digital Library. An ongoing challenge for teachers in all disciplines is identifying (and finding) high-quality online teaching resources. Anyone can post anything online—how do you know it is from a trusted source? How do you know it will work in the classroom?

The National Science Foundation (NSF), through the National Science Digital Library (NSDL), offers a mechanism by which digital learning objects can be cataloged, compiled, and disseminated. NSF is supporting the development of the Chemical Education Digital Library (ChemEd DLib), an NSDL pathway project that will make it easier for educators at all levels to find useful, tested teaching materials and help create online communities focused on special interests.

High school chemistry clubs. High School Chemistry Clubs (ChemClubs) were initiated in 2005 to nurture the interest of high school students in chemistry. Currently, there are 120 ChemClubs chartered with ACS. These clubs receive a wealth of resources from ACS, including activities related to National Chemistry Week and Chemists Celebrate Earth Day.

ChemClubs are encouraged to network with their ACS Local Section and their local Student Affiliates Chapter, thereby creating a professional continuum from high school through retirement.

The three-year pilot of this program has seen a variety of activities on the part of the ChemClubs: chem demo shows for local elementary schools, field trips, fund-raising events for those affected by Hurricane Katrina, and National Chemistry Week events. These, and other, activities enable students to experience chemistry beyond the classroom, learn about further study and careers in chemistry, and provide service to their local communities. Information about ChemClubs, including how to start a club, is available online at www.acs.org/education.

ACS textbooks. ACS produced its first textbook, *Chemistry in the Community* (*ChemCom*), in 1988. This ground-breaking text (American Chemical Society, 2006*b*) for high school students introduced chemistry on a need-to-know basis, using environmental and health issues as the context for presenting basic chemistry information. The impetus behind this, and other, ACS textbook projects is to offer a different approach to the teaching of chemistry, one that differs from traditional textbooks. Subsequent ACS texts, *Chemistry in Context* (American Chemical Society, 2006*a*) and *Chemistry* (American Chemical Society, 2005), have followed the lead of *ChemCom* by presenting chemistry in the context of societal issues using a hands-on, inquiry-based approach.

Professional development. High school teachers are valuable contributors to the American Chemical Society. Numerous teachers mentor high school ChemClubs, engage their students in the Chemistry Olympiad, organize programming at ACS national and regional meetings, and develop curricular materials for use by the chemistry community at large. These activities strengthen chemistry education by reinforcing student interest in the central science and providing professional development opportunities for teachers.

Formal professional development opportunities for high school chemistry teachers are mainly offered by ACS through workshops centered on the use of the textbooks *Chemistry in the Community* and *Chemistry*. While *Chemistry* was developed for first-year undergraduates, a handful of high school teachers are using the text for AP and IB courses. This text offers several advantages over traditional general chemistry texts by focusing on active learning and placing a greater emphasis on the biological applications of chemistry. High school teacher-leaders run the textbook-based workshops, drawing on their experiences employing these texts in their own classrooms.

Conferences and workshops. ACS National and Regional Meetings provide an important venue for high school teachers to share best practices in the classroom. High School Teacher Day programming is now fully integrated into the technical program at ACS national meetings. The Biennial Conference on Chemical Education presents another opportunity for chemical educators to interact. ACS is well represented at ChemEd conferences and at the national and regional National Science Teachers Association conferences.

Awards and recognition. Excellence in high school teaching is recognized through the James Bryant Conant Award in High School Chemistry Teaching. The George C. Pimentel Award in Chemical Education honors the accomplishments of college and university faculty members. A new national award, the ACS Award for Achievement in Research for the Teaching and Learning of Chemistry, recognizes the importance of research in identifying best practices in chemical education. These awards highlight the accomplishments of extraordinary educators who make a difference in the lives of their students every day.

Conclusion

We would all be fabulously wealthy if we could predict the future. We cannot. But we can anticipate the impact of trends in education and attempt to shape its future accordingly. More technology, increased multidisciplinarity, enhanced globalization, and a greater emphasis on standards are forces that are likely to continue to shape K–12 education, in general, and chemistry education, in particular. Exactly how these trends will play out is anybody's guess.

Louis Pasteur once noted that "In the field of observation, chance favors only the prepared mind." (Pasteur, 1854) While we cannot predict the future, we must anticipate the future, and preparing our minds for a multitude of options is our best chance for success. The American Chemical Society and its Education Division stand ready to partner with chemistry teachers to do just that.

Recommended Reading

Singer, S. R.; Hilton, M. L.; Schweingruber, H. A. Eds. *America's Lab Report: Investigations in High School Science*; National Research Council, The National Academies Press: Washington, DC, 2005. This report is not very complimentary of the status quo of lab science teaching, but does identify strengths and weaknesses, and proposes actions to take. The emphasis is on coherent units of study in which hands-on science is integrated into the learning. The report is also available online at http://www.nap.edu/catalog/11311.html.

Lechtanski, V. L. *Inquiry-Based Experiments in Chemistry*; Oxford University Press: New York, NY, 2000. This book provides 35 inquiry experiments for first-year chemistry students, including the student directions, teacher's notes, and a sample lab report.

Recommended Web Sites

The ACS Education Division offers a variety of education resources for students and teachers from kindergarten through graduate school. http://www.acs.org/education (accessed March 15, 2008).

The *Journal of Chemical Education* (JCE) provides resources that are relevant to educators in high school and higher education, including the JCE Digital Library. http://jchemed.chem.wisc.edu/ (accessed March 2008).

University of Oregon, Greener Education Materials. http://greenchem.uoregon.edu/gems.html (accessed March 15, 2008).

References

American Chemical Society (ACS). *Chemistry in the National Science Education Standards;* American Chemical Society: Washington, DC, 1997.

ACS Committee on Professional Training, *Undergraduate Professional Education in Chemistry: Guidelines and Evaluation Procedures;* American Chemical Society: Washington, DC, 2003.

ACS. Statement on Science Education Policy, http://portal.acs.org/portal/fileFetch/C/WPCP_007622/pdf/WPCP_007622.pdf, 2007 (accessed April 10, 2008).

ACS. *Chemistry;* W. H. Freeman and Company: New York, 2005.

ACS. *Chemistry in Context;* McGraw-Hill: New York, 2006a.

ACS. *Chemistry in the Community;* W. H. Freeman and Company: New York, 2006b.

Anastas, P. T.; Warner, J. C. *Green Chemistry: Theory and Practice;* Oxford University Press: Oxford, UK, 1998.

Clark, J.; MacQuarrie, D., Eds. *Handbook of Green Chemistry and Technology;* Blackwell Science: Oxford, UK, 2002.

Doxsee, K. M.; Hutchison, J. E. *Green Organic Chemistry: Strategies, Tools, and Laboratory Experiments;* Thomson Learning: Mason, OH, 2002.

Friedman, T. L. *The World Is Flat: A Brief History of the Twenty-first Century*; Farrar, Straus and Giroux: New York, 2005.

Kirchhoff, M. M.; Ryan, M. A., Eds. *Greener Approaches to Undergraduate Chemistry Experiments;* American Chemical Society: Washington, DC, 2002.

Lancaster, M. *Green Chemistry: An Introductory Text*; Royal Society of Chemistry: Cambridge, UK, 2003.

National Academies, *Rising Above the Gathering Storm: Energizing and Employing America for a Brighter Economic Future*; The National Academies Press: Washington, DC, 2007; p 6.

National Research Council, *National Science Education Standards;* National Academies Press: Washington, DC, 1996.

National Science Teachers Association, The integral role of laboratory investigations in science instruction, 2007. http://www.nsta.org/positionstatement&psid=16 (accessed March 15, 2008).

O'Connell, M. E.; Norwood, J. L., Eds. *International Education and Foreign Languages: Keys to Securing America's Future;* National Research Council, The National Academies Press: Washington, DC, 2007.

Pasteur, L. Address at the University of Lille, December 7, 1854.

Ryan, M. A.; Tinnesand, M. *Introduction to Green Chemistry*; American Chemical Society: Washington, DC, 2002.